The Grouting Handbook

The Grouting Handbook

A Step-by-Step Guide for Foundation Design and Machinery Installation

Second Edition

Donald M. Harrison

ELSEVIER

AMSTERDAM • BOSTON • HEIDELBERG • LONDON • NEW YORK • OXFORD
PARIS • SAN DIEGO • SAN FRANCISCO • SINGAPORE • SYDNEY • TOKYO

Elsevier
32 Jamestown Road, London NW1 7BY, UK
225 Wyman Street, Waltham, MA 02451, USA

First edition 2000
Second edition 2013

Notices
Knowledge and best practice in this field are constantly changing. As new research and experience broaden our understanding, changes in research methods, professional practices, or medical treatment may become necessary.

Practitioners and researchers must always rely on their own experience and knowledge in evaluating and using any information, methods, compounds, or experiments described herein. In using such information or methods they should be mindful of their own safety and the safety of others, including parties for whom they have a professional responsibility.

To the fullest extent of the law, neither the Publisher nor the authors, contributors, or editors, assume any liability for any injury and/or damage to persons or property as a matter of products liability, negligence or otherwise, or from any use or operation of any methods, products, instructions, or ideas contained in the material herein.

British Library Cataloguing-in-Publication Data
A catalogue record for this book is available from the British Library

Library of Congress Cataloging-in-Publication Data
A catalog record for this book is available from the Library of Congress

ISBN: 978-0-12-416585-4

For information on all Elsevier publications
visit our website at store.elsevier.com

This book has been manufactured using Print On Demand technology. Each copy is produced to order and is limited to black ink. The online version of this book will show color figures where appropriate.

Working together
to grow libraries in
developing countries

www.elsevier.com • www.bookaid.org

This book is dedicated to my wife Paja
who without her help this second book
would not be possible.

Contents

Preface

As with my first book on this subject the information contained within came from various sources. Some of it came from my 50-year work history in the United States, Canada, the Far East, and South America. Some of it came from conversations with others who I consider to be experts in rotating equipment and knowledgeable in the problems encountered with machinery installation.

I gathered a lot of information by attending technical conferences such as the Texas A&M Pump Symposium and the Turbo Machinery Symposium in Houston, TX. Bits and pieces were gathered in visits to engineering firms throughout the world solving specific machinery and foundation problems. I have made some mistakes and learned from them.

During my working career, I had the opportunity to work for three epoxy grout manufacturers and learned a lot from all of them.

I have not written every word in this book, and it is not my intent to plagiarize any one, far from it. I want to give credit where credit is due. The problem is a lot of my data came in the form of third- and fourth-generation copies and portions of magazine articles written by others (names unknown) that were faxed or e-mailed to me.

The intent of this book, as with the first, is to put down in writing and in a logical sequence of events, this collected knowledge so it can be used as a handy reference. I have also added numerous pictures and illustrations to better show examples of foundations, machinery support systems, anchor bolt, and grouting technology.

I once had a fellow tell me that he had 25 years of experience in machinery grouting. After observing this individual's approach to grouting for a few days, it was obvious to me that he had about 3 weeks of experience, recycled over 1,300 times.

Unlike a fine wine, improper grouting and foundations procedures do not improve with age. A bad grout job or foundation design is a bad grout job or foundation design no matter what you try to do to it later. It's like an egg, looks great on the outside, but you don't know what the inside contains.

A proper machinery installation consists of four basic factors that will determine if it will be successful and reliable over a long period of time. These are:

1. The load carrying capability of the soil the foundation will rest on.
2. The components of the foundation: mass, design, concrete mix, installation, and curing.
3. The anchor bolt and bolt design.
4. The grout.

Keep this thought in mind. The foundation and the grout hold the machine up while the anchor bolts hold it down. The epoxy grout is not a super glue to stick something to the foundation.

About the Author

Donald M. Harrison is a retired rotating equipment technician. Prior to his professional career, his military service in the United States Air Force and the U. S. Coast Guard included training in jet engine maintenance, radio communication, heavy equipment, and marine propulsion maintenance. After 12 years in the military he joined the Rotating Equipment Group at the AMOCO Refinery in Texas City, TX. Mr. Harrison has over 35 years of experience in construction and machinery installation, in the petrochemical and gas transmission industries. He spent 14 years working for three major grout manufacturers (Ceilcote, Master Builders, and ITW Polymer Technologies (Formally Philadelphia Resins)). He has authored numerous papers on reciprocating compressors and their components, grouting and foundations.

Don lives in Santa Fe, TX with his wife and enjoys building Hot Rods and restoring Antique Cars with his grandson Christian. He is available for consultation on foundation and grouting problems, and may be contacted at: donh2013@aol.com.

Acknowledgments

Special thanks to:

Richard O'Malley
World Grout Manager
ITW Polymer Technologies
Montgomeryville, PA.
Romalley@itwpolytech.com
Richard was invaluable in providing pictures, drawings, and epoxy grout technical data for this project.

Mike Smith
Dresser Rand
Louisiana, MO.
mdsmith@dresser-rand.com
Mike provided information and pictures for ®Magbolt and ®Vibratherm chock.

Mark Krenek
Technical Consultant
Reynolds-French & Co.
Tulsa, OK.
Mark@r-f.com
Mark provided technical info on ®Vibratherm.

Eric Harrison
Field Service Manager
Euro Gas Systems
Mures Romania
eric.harrison@eurogassystems.com
Eric provided assistance with portions of the Chocking technology and skid grouting.

Jason Milburn
Marketing Manager
Superbolt (part of the Nord-Lock Group)
1000 Gregg St.
Carnegie, PA 15106
P: 412-279-1149
bolting@superbolt.com
www.superbolt.com • www.nord-lock.com

1 The Foundation

Your Machinery reliability will be directly proportional to your investment in a properly designed and installed foundation. Shortcuts or poor practices taken here will directly affect rotating equipment life.

The Initial Machinery Foundation Design

We spend a lot of engineering time and money designing a foundation for a piece of equipment. Contrary to popular belief, machinery foundation structural design is still evolving. We must rethink today's foundations to eliminate the design stress concentration points to reduce the possibility of a crack developing in the concrete. The most common causes of cracking in large concrete foundations are inside right angles (Figure 1.1) built into the foundations. You can easily eliminate these stresses by incorporating a chamfer of 3–6 in. into the design.

The Grouting Handbook. DOI: http://dx.doi.org/10.1016/B978-0-12-416585-4.00001-8

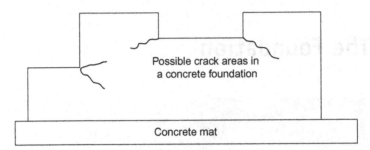

Figure 1.1 Depending on the dynamics of the machinery installed, this type of cracking can be caused by the right angle stress concentration points originally designed into the foundation, and by the thermal growth of the machine.

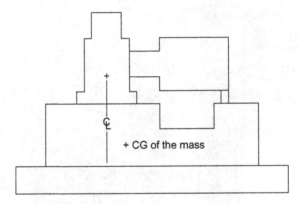

Figure 1.2 For reciprocating engines and compressors, the centerline of the crankshaft should be within 6–8 in. of the center of gravity of the total foundation mass.

Reciprocating Compressor and Engine Foundations

Machinery foundation structural design is *still* evolving. Even today, we do not give enough thought or research to designing a foundation for the long term. The foundation block width should be greater than its height to overcome the cantilever effect experienced by most reciprocating engines or compressors. A good rule of thumb for engine and compressor foundation design is to keep the blocks low and wide (Figure 1.2). The center of gravity of the foundation's total mass (block and mat) should be within 6–8 in. of the vertical centerline of the crankshaft.

Another rule of thumb calls for the mass of a machinery foundation for a reciprocating compressor and driver to be, at *minimum*, five times the combined weight of those items. In some cases, mass ratios may go as high as eight times the machine weight. This ratio depends on the unbalanced forces generated by the machine.

Most original equipment manufacturers (OEMs) typically will supply a drawing that specifies a generic foundation design for their machine. This drawing usually has a disclaimer saying something like:

The XYZ Company offers this design only to show locations of foundation bolts and general dimension recommendations where soil conditions are satisfactory. XYZ assumes no liability whatsoever for the foundation design. Proper concrete mixture, correct reinforcement, sufficient mass, and satisfactory footing is essential to give permanent support and prevent vibrations. If local soil conditions are poor, a competent foundation engineer should be employed.

Skid-Mounted Equipment Foundations

As mentioned earlier, the general rule of thumb for reciprocating equipment foundation design is for the foundation to be a minimum of five times the mass of the operating equipment. For example, if a reciprocating engine and high-speed compressor assembly weighed 25,000 lbs, the underlying concrete mass should weigh about 125,000 lbs. Sometimes, installing a foundation of this mass for a reciprocating skid-mounted assembly is not practical or even feasible. To compensate for this lack of foundation mass, you sometimes can fill the individual skid compartments with concrete at the time of fabrication. The idea is for this concrete to act as a damping agent by adding mass to the skid. The concrete is added after the skid has been fabricated and painted. Unless some type of reinforcing steel is welded to the inside of the individual skid compartments, the concrete will not perform as expected because it will not bond to the painted steel. If this occurs, its damping characteristics are reduced.

In general, the foundation mass for skid-mounted equipment installations of this type need to be a minimum of 1½ times the weight of the total completed skid package (Figure 1.3).

On some skid-mounted equipment, the natural frequency of the skids has been measured at 200 Hz or more. The natural frequency of the skid is totally dependent on the skid design and the equipment operating frequency.

Foundations for Centrifugal Pumps

Over the years, several papers and articles have been written on epoxy grouting of American Petroleum Institute (API) and American National Standards Institute (ANSI) pumps. The generally accepted rule for pump foundations is that they should be three times the mass of the pump, driver, and steel baseplate. Other than that, not much has been written about the basic foundation dimensions for API and ANSI pumps.

Many years ago, Ray Dodd of Chevron's Pascagoula, Mississippi facility developed a rule of thumb concerning the sizing of pump foundations (Figure 1.4). He found a

Figure 1.3 Skid-mounted compressor unit.

way to determine if pump foundation size is adequate. If the lines pass through the bottom of the foundation, the foundation is adequate. If the lines pass through the sides, then the foundation is too narrow.

He arrived at this rule of thumb from a blend of engineering principles and his experience in the field of rotating equipment. His intent was to provide a fast yet simple method for the *evaluation* of existing or future pump foundations, something that could also be used to supplement sketchy general installation data from the pump OEM. Additionally, Ray found that this rule of thumb was also an excellent way to verify foundation designs furnished by the pump manufacturers.

Today's pumps differ in their design and construction but retain the same basic dynamic principles that apply to all of them. You must consider both the dynamic and static forces of the pump and its foundation for adequate long-term soil support.

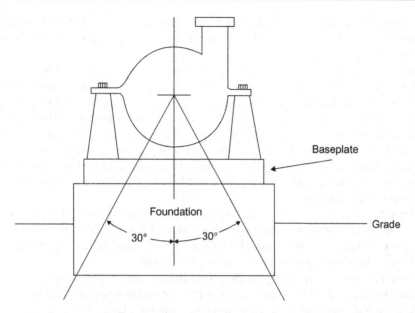

Figure 1.4 Lines drawn 30° down from the shaft centerline are a good rule of thumb to show if pump foundation is wide enough.

According to Ray Dodd's rule of thumb, a well designed pump foundation and steel pump base allows imaginary lines extending downward at 30°, from either side of the vertical through the pump shaft, to pass through the bottom of the foundation and not the sides.

The following guidelines will apply to good foundation design for pumps:

- The mass of the concrete foundation should be a minimum of three times the mass of the supported equipment and should have sufficient rigidity to withstand the axial, transverse, and torsional loadings generated by these machines.
- The foundation should be 6 in. wider than the baseplate for pumps up to 500 hp and 10 in. wider for larger machines.
- The concrete used in the foundation should have a *minimum* tensile strength of 350 psi.
- Epoxy grout should always be used to mate the pump baseplate to the foundation.

These guidelines and Ray Dodd's rule of thumb can be used as a quick evaluation of a suspected pump foundation as potential sources of vibration.

Most pump foundation vibrations are usually observed at rotational speed. If you identify a foundation as the source of a vibration problem, it may be due to inadequate mass or configuration. Adding concrete mass to an existing foundation may not be the answer. If you add a concrete mass equal to the full weight of the pump and its steel baseplate to the existing foundation of a machine, the resulting velocity energy will be approximately half that of the original design, providing that the *cold joint* can be eliminated by using an epoxy bonding agent or by later pressure-injecting the cold joint with epoxy.

Adding twice the mass of the pump and baseplate to an existing foundation reduces the vibration energy to about ⅓ of the amount experienced on the original foundation design without the added mass.

W.E. Nelson, P.E., was a turbo machinery consultant in Dickinson, TX. I worked with him in one of the largest refineries in the world, and he was my mentor. He said that "his experience with adding additional concrete mass to an existing pump's foundations was less than satisfactory." He found that, of the numerous existing pump foundations that received this additional mass of concrete, only 20% performed as expected. The remaining 80% showed no appreciable improvement in the reduction of vibration. Additionally, he found that for the same cost as the additional concrete for mass, associated rebar, and doweling, he could have replaced the existing foundation with a correctly sized one.

Perry Monroe, P.E. (retired turbo machinery consultant in Livingston, TX), cautions that "a cold joint will develop when adding additional concrete mass to an existing foundation. Even the use of steel dowels will not eliminate the effects of a cold joint where the new concrete meets the old concrete."

Both men advised "spending the money up-front to get an adequately sized foundation." Failure to do this could result in expensive maintenance to reengineer bearings, seals, and couplings in an effort to reduce vibration levels that were originally aggravated by an undersized foundation.

Looking at another aspect of the foundation mass problem, many pumps in the 5–50 hp range are installed without any forethought given to specially constructed foundations. These foundations are normally just large enough for the pump and its baseplate. If you are dealing with an existing pump foundation and want a quick and cost-effective fix, add the new mass using a low-exotherm, deep-pour epoxy grout instead of concrete. Epoxy grout will achieve a better bond to the existing concrete than new concrete. Additionally, the cure time for most low-exotherm epoxy grout is hours instead of days for concrete.

Cured concrete weighs about 150 lb/ft^3 and, when compared to steel, has a vibration damping capability six times greater than steel (Figure 1.5). Epoxy grout weighs about 130 lb/ft^3 and has a vibration damping capability that is 26 times better than

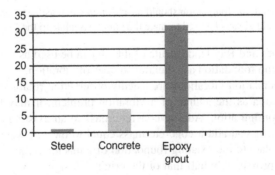

Figure 1.5 A comparison of the vibration damping capability of steel, concrete, and a well-known red epoxy grout.

concrete. Also, the epoxy grout will develop a bond to old concrete stronger than the concrete tensile strength and will develop a bond to properly prepared steel in excess of 2,000 psi.

Regardless of whether the unit is an API or ANSI pump, to effectively dampen and dissipate vibration, two things must occur:

1. You must correctly size the foundation.
2. The foundation, steel baseplate, and the pump must be a monolithic unit. You can accomplish this only during installation by using epoxy grout and proper grouting procedures.

Foundation Mass to Equipment Weight Ratio for Pumps

A generally accepted rule of thumb calls for the *minimum* mass of a pump foundation to be three times the weight of the machine and driver, depending on the operating dynamics. In some cases, this figure needs to be four to five times the weight of the equipment.

Soil Conditions

Proper soil compaction is an absolute must because excavating for a foundation loosens the soil. This loosened soil will, over time, allow the foundation to settle. When testing for soil compaction and density, take several readings in different locations, not just one.

After excavation, it is extremely important to stabilize the soil before pouring a machinery foundation. Most people will use a compatible fill that is locally available and you hope will provide the necessary load carrying capability to properly support the foundation, machine, and associated piping. After compaction, this fill will normally be covered with a concrete mat before the foundation is poured. The mat thickness will vary depending on the foundation design, soil density, and the machine dynamics. The total load forces applied by the mat to the compacted soil must be uniform. You must give forethought to design considerations at this point.

The operating machine frequency should not be greater than 0.8–1.2 times the natural frequency of the soil. The damping coefficient of soil can range from 0.03–40%.

Depending on the size of the machine you are installing, you may want to have a soil exploration test made to a depth of at least 8–10 ft (for up to 200 hp units). Testing laboratories will then determine the soil bearing pressure ability (they will measure two or three points to be certain) of the strata at foundation support level. This gives you the allowable static soil loading in pound per inch. You can determine the *allowable soil dynamic load* by dividing the static load by four. As a preliminary check, add the weight of the machine and foundation block and divide by the dynamic soil load to determine the required area in square feet for the bottom surface of the block. If this area is greater than the area shown in the vendor's foundation drawing, you have to add a mat to support the block and to distribute the mass over the larger area. *A piling plus mat extension may be called for if the soil tests show a water table above the support level.*

Table 1.1 Soil Static and Dynamic Load Values

Type of soil	Soil Static Load (Lb/ft^2)	Safe dynamic Load (Lb/ft^2)
Rock	8,000	2,000
Dry Gravel	4,000	1,000
Dry Coarse Sand	4,000	1,000
Dry Fine Sand	3,000	750
Firm Clay	2,000	625
Soft Clay	2,000	500
Wet Clay	1,000	250
Wet Sand & Clay	880	220
Alluvial Soil	800	200

The importance of soil base tests is easy to understand when you consider that the vendor's drawing is usually predicated on dry, firm soil of between 1,500 and 2,000 lb/ft. The chart in Table 1.1 gives the average (tested) values of various soils.

The dynamic conditions for a foundation are machine speed and operating conditions or volume. If the dynamic operating conditions of a foundation are changed by as little as 10%, foundation problems could result.

You must isolate the foundation as much as possible to prevent the transmission of vibration to adjacent buildings or structures. An effective procedure is to surround the foundation with a coarse gravel or a 50/50 mix of gravel and sand (not fill dirt) when backfilling. Also, you should isolate all floor slab or area paving from the foundation block with some type of flexible and waterproof isolation material.

Proper soil compaction is an absolute must because excavating for a foundation loosens the soil. This loosened soil will, over time, allow the foundation to settle. When testing for soil compaction and density, take several readings in different locations, not just one.

Due to foundation settling over time, the resultant piping strain on the suction and discharge nozzles of the pump can be directly related to alignment problems of the equipment.

Concrete for the Foundation

The epoxy grout is the bond between the equipment and the foundation and probably the most important aspect of a unitized installation. For epoxy grout to do its job, the concrete mix design must have a high tensile strength because a low tensile strength may cause the epoxy to delaminate from the concrete surface during the cure period. This delamination is called *edge lifting* or *curling* and will be addressed in more detail later in this book. Edge lifting is often the result of weak concrete, usually caused by water added at the job site to facilitate concrete placement or insufficient surface preparation prior to grouting.

Figure 1.6 Do not allow water to be added randomly to the mix.

Adding Water to the Concrete Mix at the Job Site

When you are pouring epoxy grout on a concrete foundation, it is important that the concrete has a tensile strength of no less than 350 psi. Usually, concrete is specified by its compressive strength, such as 3,000 or 3,500 psi. Unfortunately, high compressive strength concrete does not necessarily result in a high tensile strength. However, a high tensile strength in concrete will result in high compressive strength.

Weak concrete can originate at two locations: the concrete batching plant or the job site. Very seldom does the end user have a qualified inspector at the batch plant to test and witness the actual batching of the concrete. To exacerbate this problem, no specifications may dictate how the concrete will be handled as it is transported from the batch plant to the job site.

At the job site, the truck operator or the cement finisher will be the determining authority on how much water to add to make the concrete flowable and workable for their purposes (Figure 1.6).

Adding 1 gal of water to each cubic yard of 3,000 psi concrete results in the following:

- The slump is increased by about 1 in. (Concrete slump should be between 3 and 5 in.)
- The compressive strength is reduced by about 200 psi.
- The tensile strength is reduced by about 20 psi. (Concrete tensile strength is about 10% of its compressive strength.)
- The effective loss of one-fourth of a bag of cement is caused.
- The shrinkage (cracking) potential of the concrete is increased.

Specialized concrete formulations with improved flow characteristics are now available on the market. They require less water and less vibrating, which reduces the chances of aggregate settling. With some of these formulations, shrinkage can be reduced in the early stages of the cure. During hot weather, the concrete setting time will be shorter. The mixer at the job site may be tempted to add more water to extend its working time. This is called *re-tempering* and can be extremely detrimental to the strength of the concrete. You can control concrete setting time and water requirements by using concrete with admixtures that reduce the water requirements and retard or, in some cases accelerate, the set time. Using these admixtures results in a concrete that can maintain a higher slump for a longer period of time and will requires less water.

Some other benefits for using admixtures are:

- Less heat from friction developed in the concrete truck during mixing;
- Extended life during transport and placement;
- Improved handling characteristics and physical properties.

What Are Concrete Admixtures?

Admixtures are ingredients other than water, aggregates, hydraulic cement, and fibers that are added to the concrete batch immediately before or during mixing.

The proper use of admixtures offers beneficial effects to concrete, including improved quality, acceleration or retardation of setting time, enhanced frost and sulfate resistance, control of strength development, improved workability, and enhanced finishability. It is estimated that 80% of concrete produced in North America these days contains one or more type of admixtures. According to a survey by the National Ready Mix Concrete Association, 39% of all ready mixed concrete producers use fly ash, and at least 70% of produced concrete contains a water reducer admixture.

Admixtures vary widely in chemical composition and many perform more than one function. Two basic types of admixtures are available: chemical and mineral. All admixtures to be used in major concrete construction should meet specifications: tests should be made to evaluate how the admixture will affect the properties of the concrete to be made with the specified job materials, under the anticipated ambient conditions, and by the anticipated construction procedures.

Mineral Admixtures

Mineral admixtures (fly ash, silica fume (SF), and slags) are usually added to concrete in larger amounts to enhance the workability of fresh concrete; to improve resistance of concrete to thermal cracking, alkali-aggregate expansion, and sulfate attack; and to enable a reduction in cement content.

Fly ash
Silica fume
Ground granulated blast furnace slag.

Chemical Admixtures

Pouring fresh concrete is a time-sensitive project and unexpected delays can cause major problems. With the use of admixtures, you can have more control over your concrete. Admixtures can restore loads of concrete that might need to be rejected due to delays or other complications. They can improve the performance of problem concrete by modifying its characteristics and enhancing its workability.

Chemical admixtures are used to improve the quality of concrete during mixing, transporting, placement, and curing. They are added to concrete in very small amounts mainly for the entrainment of air, reduction of water or cement content, plasticization of fresh concrete mixtures, or control of setting time. Specialty admixtures can include corrosion inhibitors, shrinkage control, and alkali-silica reactivity inhibitors.

There are seven types of chemical admixtures specified in American Society for Testing and Materials (ASTM) C 494, and AASHTO M 194, depending on their purpose or purposes. Air entraining admixtures are specified in ASTM C 260 and AASHTO M 154. General and physical requirements for each type of admixture are included in the above-mentioned specifications.

Air entrainment
Water reducing
Set retarding
Accelerating
Super plasticizers.

ASTM C494 specifies the requirements for seven chemical admixture types. They are:

- Type A: Water reducing
- Type B: Retarding
- Type C: Accelerating
- Type D: Water reducing and retarding
- Type E: Water reducing and accelerating
- Type F: Water reducing, high range admixtures
- Type G: Water reducing, high range, and retarding admixtures.

NOTE: Changes occur in the admixture industry faster than the ASTM consensus process. Shrinkage reducing admixtures (SRA) and midrange water reducers (MRWD) are two areas for which no ASTM C494-98 specifications currently exist.

Admixture Types

Admixtures are normally provided as water-based solutions and can be added to the concrete at up to 5% on cement weight, although most types are added at less than 2.0% and the majority is at less than 1.0%.

The types of admixture are as follows:

Normal water reducing/plasticizing admixtures: Used to increase the workability of concrete at a consistent water content and/or reduce water by up to 10%. These are used by most ready mix companies to optimize the concrete performance for normal concrete.

High range water reducing/superplasticizing: These admixtures give a much higher performance than the normal plasticizers. It is used to give very high levels of workability or water reductions ranging from 12% to over 30%. They can be used extensively on larger projects where reinforcing steel requires high workability of the concrete and can be used in precast and on site where the large water reduction provides very high early strength and improved durability. They are essential to the production of self-compacting concrete (SCC).

Retarding: These admixtures slow the rate of cement hydration, preventing the cement from setting before it can be placed and compacted.

This type of admixture is mainly used in hot conditions and climates or on very large pours.

Accelerating: Admixtures used to increase the rate of early hydration of the cement.

This product can accelerate the setting or the early strength development. Accelerating admixtures are used mainly in cold conditions.

Air entraining: These admixtures cause tiny air bubbles <0.3 mm in diameter to stabilize within the cement paste.

This air helps to prevent the concrete from cracking and scaling as a result of frost action. Air also increases cohesion in the mix, reducing bleed water and segregation of the aggregate before the concrete can set.

Water resisting (waterproofing): These water repellent admixtures block or impede the flow of water through the natural capillaries in hardened concrete.

Used in structures below the water table or in water-retaining structures.

Corrosion inhibiting: These admixtures work for many years after the concrete has set, increasing the corrosion resistance of reinforcing steel to reduce the risk of rusting steel causing the concrete to crack and scale.

Segregation reducing, viscosity modifying admixtures: SCC uses super plasticizers to give a low yield, highly fluid mix but may also require a segregation control or viscosity modifying admixture (VMA) to ensure that mix cohesion is maintained.

Shrinkage reducing admixtures: Concrete shrinks, mainly due to loss of excess water. This causes internal stresses that lead to cracking or curling, especially in slabs. These admixtures reduce the shrinkage stress.

Manpower

It is extremely important to have sufficient crew and equipment on hand to successfully place and work the concrete. When the concrete trucks arrive, the crew

Figure 1.7 Plan on some type of a delay.

should be ready to discharge their loads as soon as possible. If a delay occurs, do not let the trucks stand by without continuing to rotate the mixer *slowly*. Prolonged mixing in hot weather increases the temperature of the concrete and causes faster hydration. This accelerated hydration rate will shorten the time available for placing the concrete.

Should problems arise at the job site that will delay the placement of the concrete, promptly notify the ready mix plant so that they can reschedule the batching of your order and avoid standing by with, or waiting on, your deliveries.

Transport time from the batch plant to the job site (regardless of the ambient temperature) should not be longer than 45 min to 1 h unless special procedures or additives are used. Plan for a delay of some type, as shown in Figure 1.7, you never know what will happen.

Vibrate the fresh concrete without delay but do not overvibrate it. Overvibrating can cause excessive settling of the aggregates in the mix and increase the amount of laitenance after the concrete hardens.

After surface finishing, apply a curing membrane to prevent rapid moisture evaporation. The foundation should then be covered with a material such as polyethylene sheeting as shown in Figure 1.8.

To counteract the effects of high ambient temperatures, low humidity, or wind, you may need to erect some type of environmental control to shade or protect the fresh concrete, prevent stiffening, and minimize plastic shrinkage.

NOTE: Epoxy grout will develop a bond to concrete greater than the tensile strength of the concrete on which it is poured.

Figure 1.8 Foundation covered with polyethylene sheeting to retard rapid evaporation.

The Concrete Mix

Fatigue in concrete is accumulative, and vertical cracks located in the midsection of a high dynamically loaded compressor foundation are usually caused by torsional problems. Most machinery foundation designs do not have sufficient reinforcing steel in them. The minimum amount of reinforcing steel used in a foundation should be no less than 2% of the volume of the foundation. For reciprocating foundations with unbalanced forces acting upon them, you should use transverse rebar.

The best cement to use for machinery foundations is Type I or II. This is addressed later in this chapter in section "Types of Portland Cement." Adding fly ash to the concrete improves its quality by lowering the heat of hydration, increasing the density, and reducing the permeability. Use crushed stone instead of round river rock because the crushed stone provides a better surface profile for the cement to bond to than the smooth surface of river rock.

Most engineering specifications simply call for a 3,000–3,500 psi concrete mix. Little or no thought is given to the tensile strength of the concrete. It is a generally accepted rule of thumb that the tensile strength of concrete is 10% of its compressive strength. Most concrete testing is for compressive strength; very seldom is the tensile strength tested. Most tests done for tensile strength reveal that, although the compressive strength is adequate, the tensile strength is very low, usually below 275 psi. Epoxy grout needs a minimum of 350 psi tensile strength concrete to reduce the possibility of edge lifting. Unfortunately, when it comes to the actual installation of the foundation, most of the engineering and design work is wasted because the specific design of the concrete mix is left up to the batch plant.

Very seldom does the end user have a qualified inspector on site *at the batch plant* to test and witness the actual batching of the concrete. To further exacerbate this problem, there may be no specifications dictating how the concrete will be handled from the batch plant to the job site. At the job site, the driver of the concrete truck and the cement finisher determine how much water must be added to make the concrete flowable and workable. After the concrete is poured, little thought is given to proper curing of the foundation.

Consider the following when selecting concrete:

- High tensile strength concrete *does* result in high compressive strength.
- High compressive strength concrete *does not* necessarily result in high tensile strength.
- Aggregate should be crushed stone.

During the 1950s and 1960s, foundations were installed that were adequate by the engineering standards for the machine dynamics at that time, but are grossly underdesigned in mass and rigidity when today's technology is applied. Fifty years later, we are taking these foundations out and replacing them. Are we replacing them with foundations commensurate with 2012 technology?

Maybe not—this is what we have learned:

- Compared to replacement or repair cost, rebar and proper design are the least expensive part of a foundation.
- Acceptable concrete is a must. Obtain the services of a good civil engineer or foundation consultant who understands machinery foundations to write a concrete specification, if your company does not already have one.
- The installer must be able to follow the design engineer's specifications. Employ a contractor who has experience in pouring large foundations, not just one who submits the lowest bid.
- Available real estate in existing operating units governs foundation size in many instances.

Specialized concrete formulations with improved flow and slump characteristics are now available in the market. These formulations use concrete admixtures, such as super plasticizers, fly ash, and other special chemicals. The slump requirement for regular concrete is 3–5 in. The slump for these specialized formulations can be anywhere from 6–9 in. Because the use of super plasticizers improves the flow, the mix requires less water and less vibrating. This will also reduce the chances of aggregate settling in the concrete mix. With some of these formulations, you can reduce shrinkage in the early stages of the cure, allowing epoxies to be poured within 72 h.

In an effort to address certain aspects of the repair market, polymer-modified concrete repair compounds have been introduced to reconstruct the upper portions of large and small foundations when a rapid cure is required. Their manufacturers claim these compounds are more cost-effective, but with a price of about $180 to $190/ft^3, they are only slightly lower in cost than epoxy grout at approximately $200/ft^3. The anticipated savings are only a few dollars per cubic foot, and these compounds still require you to come back and pour an epoxy grout under the equipment being grouted. At this point in time, no one can say there is anything wrong with the polymer-modified concretes. There have been as many good reports as bad reports about them. Unfortunately, no long-term history on these polymer-modified concretes exists, so we do not know what their performance will be 10 years or more in the future. You may draw your own conclusions as to the supposed cost savings based on the installation comparison found in Chapter 3.

Placing Concrete in Hot Weather

Concrete setting time decreases more quickly in hotter weather; therefore, its consistency will thicken. As a result, the mixing crew will often be tempted to add more water at the job site. Concrete setting time and water requirements can be controlled by using concrete containing admixtures that reduce the water requirements and retard (or accelerate) the set time. The result of these admixtures is that the concrete can maintain a higher slump for a longer period of time and requires less water. Some other benefits to using admixtures are:

- Less heat from friction developed in the concrete during mixing
- Less heat from its chemical cure developed in the concrete
- Extended life during transport and placement
- Improved handling characteristics.

It is extremely important to have a sufficient crew, materials, and equipment on hand to successfully place and work the concrete. If you occasionally spray the ground, rebar, and forms with water, you can prevent water absorption from the concrete into the ground. To counteract the effects of high ambient temperatures, low humidity, or wind, you need to erect some type of environmental control to shade or protect the fresh concrete, prevent stiffening, and minimize plastic shrinkage.

If problems arise at the job site that will delay the placement of the concrete, promptly notify the ready mix plant so that they can reschedule the batching of your order and avoid standing by with your deliveries. When the ready mix trucks arrive, be ready to discharge their loads as soon as possible.

CAUTION: Prolonged mixing in hot weather increases the temperature of the concrete and makes it hydrate more quickly. This accelerated hydration rate will shorten the available time for placing the concrete.

NOTE: Transport time from the batch plant to the job site regardless of the ambient temperature should not be longer than 45 min to 1 h unless you use special procedures or additives.

Placing Concrete in Cold Weather

The term *cold weather* refers to ambient temperatures from 55°F to 33°F. At these lower temperatures, concrete set time and strength development is delayed. To counteract the effects of cold weather concreting, you can use the following methods:

- Heat the concrete mix water.
- Heat the concrete materials.
- Control the area temperature where the concrete is to be placed. This means raising the temperature of all surfaces that will be in contact with the new concrete (rebar, forms, subgrade).
- Use additional cold weather admixtures in the mix.

Curing protection for cold weather should be continuous from start to finish and should not be interrupted until the concrete has reached its designed strength. Rapid surface drying of the concrete must be avoided.

For every 20°F drop in concrete temperature, the set time will be doubled.

Concrete Curing for Machinery Foundations

After the concrete is poured, little thought is given to proper curing of the foundation. How fast water hydrates from the fresh concrete will affect the ultimate concrete strength.

Improperly cured concrete can have its designed strength reduced by as much as 50%. The term *curing* simply means retaining the concrete in a protected environment long enough for all components to chemically combine with the cement to form a tough adhesive that results in a strong, durable concrete. Good curing also means keeping the concrete damp and at a uniform temperature until it reaches its desired compressive strength. All concrete must be properly cured to achieve its maximum strength. Properly cured concrete is superior to improperly cured concrete because it shrinks less and is more resistant to cracking.

The curing process should start as soon as the concrete is hard. This means hours, not days, after the concrete is poured. Early drying in hot, windy weather must be prevented. Good concrete, properly cured, will have fewer pores and crevices. The concrete will be more durable and less prone to dusting, cracking, crazing, or spalling. In general, the better the cure, the better the concrete.

Because concrete develops its strength more slowly at temperatures below 40°F, do not expect the concrete to perform satisfactorily unless you cure it properly. You must protect the concrete from temperature extremes and not use it until it has developed the required strength. You must erect suitable structures to protect the fresh concrete and control the environment of the area.

Methods for Curing Concrete

The most common curing compounds are the *membrane curing compounds*. You may apply these by spray (Figure 1.9), brush, or roller, and they are relatively low in

Figure 1.9 Spray-applied curing membrane.

cost. You can use two kinds of chemicals as *curing membranes* for concrete: water-based and oil-based. These membranes form a temporary seal that prevents rapid water loss during cure on new concrete. Water-based is preferred because it can be washed off after the concrete is cured. Oil-based, however, must be acid-etched, sandblasted, or chipped to remove.

- Periodic water spray is not a good curing method. Proper curing requires the concrete to be kept continuously damp. Allowing the concrete surface to dry between spraying can cause surface crazing and cracking of the slab.
- Waterproof curing paper holds moisture and temperature in the concrete. To use this material, you must wet the concrete first and then cover it with the waterproof paper.
- Damp burlap is a method of curing usually associated with highway construction. It often leaves a discoloration in the concrete.
- Plastic sheeting (Figure 1.10) is completely watertight, lightweight, and easy to handle. However, the sheeting must lie flat against the concrete surfaces to be completely effective. You can accomplish this by spreading sand over the sheeting after it has been applied.
- Water flooding (Figure 1.11) of a concrete foundation will prevent rapid evaporation and cover all surfaces (vertical and horizontal) of the foundation. This is rather labor intensive because someone must monitor the water flood around the clock.

Whatever method of curing you employ, you should keep the concrete moist for a *minimum* of 7 days.

CAUTION: Curing membranes should not be regarded as primers for epoxy coatings or epoxy grout and must be completely removed before applying either of these products. Removal of curing membranes can be done at when preparing foundation for grouting.

Figure 1.10 Polyethylene sheeting is applied to reduce rapid drying.

Figure 1.11 Water flood is another form of reducing the possibility of rapid drying but is labor intensive.

Types of Portland Cement

There are eight types of cement. The *type* of cement depends on how finely the cement is ground.

Type I: Referred to as normal, regular, or standard cement. It has a 95% pass using a 325 sieve and will provide 3,000–4,000 psi concrete in 28 days, depending on the environment, amount of water used, and type of aggregate. This is normally used for general construction (house foundations, sidewalks, driveways, etc.)

Type IA: Same as Type I, but air entraining.

Type IP: Blended cement also used in general construction. It contains Portland cement and fly ash and has a lower rate of hydration. It will take longer to gain its full strength (28 days plus). The ultimate strength will be lower than Type I. Normally used for massive structures (dams, piers, and so on).

Fly ash is used to:

- Increase material strength
- Decrease the number of pores (make the concrete more dense, decrease its permeability)
- Reduce cost.

Type II: Moderately corrosion-resistant, contains less tricalcium aluminate, and is used where sulfates or H_2S is present. Type II will usually generate less heat at a slower rate than Type I. Normally used for massive pours (machinery foundations). Using it will reduce temperature rise, which is especially important in hot weather.

Type IIA Same as Type II, but air entraining.

Type III: Generally referred to as *high early*. It is used where forms need to be removed as soon as possible, or where the structure must be put in service quickly. In cold weather, using it permits a reduction in the controlled curing period. Chemically the same as Type I, but ground much finer (100% pass using a 325 sieve). Richer mixes of Type I can be used to gain high early strength, but Type III may provide it more satisfactorily and economically.

Type IV: Referred to as *low heat*. It is cement manufactured for the building of large structures, such as the Hoover dam. It has an extremely low rate of hydration. However, it has not been used in recent years because it has been found that in most cases, heat development can be satisfactorily controlled by other measures.

Type V: Used only where the concrete will be exposed to severe sulfate attack. Chemically the same as Type II, but contains even less tricalcium aluminate. It is more sulfate-resistant than Type II and gains strength more slowly than Type I.

As the numbers go up (Type I, Type II, and so on), the amount of cement fines in the mix increases. As the amount of cement fines increases, the mix becomes more unstable due to the increased exothermic reaction of the cement. Many a concrete pour has gone bad in cold weather because someone tried to compensate for the cold by using a 10-sack mix without knowing what the effect would be. When pouring machinery foundations, someone with experience in the application of high-strength concrete and in mix design and batching procedures must be at the job site.

Contamination of the Foundation

It is not uncommon for an otherwise good grout, soleplate, or chock installation to detach itself from the foundation after time due to oil penetration into the concrete. Oils, acids, and other chemicals can weaken cementious materials quickly and cause them to fail in compression, tension, or shear. Whatever grout or chock system you use, it must be protected by painting all absorbent surfaces with at least two coats of good quality epoxy paint. Any cracks that exist or develop subsequently must be sealed. Because, in the past, no one gave much thought to protecting concrete, today it is rare to find an old machinery foundation without any cracks or contaminated concrete. Remember that the painting of the exposed areas of a new foundation with epoxy paint will prevent sulfate attack and the formation of alkali-silica gel.

If the concrete is cracked but clean, epoxy resin injection can bond the concrete together again. Any oil contamination of the cracked areas will make this repair very unlikely to succeed. Where cracks are likely to be a structural weakness or are seen to "work," posttensioning rods should be considered. All cracks should be sealed at the surface to prevent oil, water, and such from penetrating. If any repairs to old foundations are to be made with concrete or cement, always use a bonding adhesive between the new and old concrete materials. A sound foundation block is essential for long-term and successful machine operations.

Guidelines for Use of Epoxy Grout on Oil-Saturated Concrete

The objective when using epoxy grout or any epoxy material on oil-saturated concrete is to produce results comparable to the properties of good concrete because these were the design criteria used in the original installation design.

Concrete can absorb oil but, fortunately, the process is relatively slow. After oil has been absorbed, a gradual degradation in both tensile and compressive strength will follow. Given enough time, concrete can become so weak that it can be crumbled between the fingers. Preventative measures during construction, such as sealing the foundation with two coats of epoxy paint to provide an oil barrier, will prevent this problem. You can use remedial techniques after oil degradation has occurred; most involve either removal or replacement.

The extent to which you must prepare the surface of oil-saturated concrete before applying epoxy grout depends largely on the following factors:

- The thickness at which the epoxy grout will be applied
- The tensile strength of the contaminated concrete
- The type of forces to which the foundation will be subjected (dynamic, tensile, or torsional).

Epoxy grout thickness should vary inversely with the strength of the underlying concrete. In solid materials, forces resulting from compressive loading are dispersed throughout the solid in a cone-shaped pattern, with the apex at the loading point. In tensile point loading, the force pattern is such that, on failure, a hemisphere-shaped crater remains. Consequently, the weaker the concrete, the thicker the epoxy covering should be to disperse the loading sufficiently before the force is transmitted to the concrete.

There have been many repair jobs done with epoxy on foundations of reciprocating gas engine compressors with such widespread oil degradation that it was impossible to remove all oil-soaked concrete before regrouting. In these cases, the majority of the regrouting was carried out with the equipment in place. These repairs were accomplished by chipping away the old grout and contaminated concrete from under the machine. The weight of the machine is usually supported by the equipment leveling screws, or temporary jack screws on elevated supports made from steel pipe or other solid material as shown in Figure 1.12.

After most of the contaminated concrete is removed, drill or core drill 1½–2 in. vertical holes into the remaining contaminated concrete. These holes should be on 18–24 in. centers and, if required, should extend deep enough to pass through the remaining contaminated concrete into at least 2 ft of solid concrete. Depending on the dynamics of the machine and the strength of the concrete, you may need to angle-drill additional holes in the outer periphery of the foundation to cross at an elevation of 2–3 ft below the bottom of the existing concrete surface. Grout in 1–1½ in. reinforcing steel into the vertical and angled holes with a liquid or flowable epoxy.

Additional horizontal reinforcing steel may be wired to the newly placed vertical and angled steel. The purpose of this new reinforcing steel is to transfer as much load as possible into the solid portion of the foundation.

Allow the epoxy enough time to cure and proceed with reconstituting the foundation. If epoxy grout is to be used, refer to Chapters 4 and 8.

Figure 1.12 Machinery supported by supports fabricated from pipe.

What Is the Acceptable Moisture Content in Concrete? Or, How Soon Can We Pour the Epoxy Grout on the New Concrete?

"How long does it take concrete to cure?" This depends on who you talk to. On average, it takes concrete about 50 years to properly cure; it lies dormant for about 1,000 years; and then it starts to decay ... we think. This all depends on how the concrete was cured.

The question most asked of grout manufacturers is "How long do we have to wait after the concrete is poured before we can apply the epoxy grout?" The question that should be asked is "What is the percentage of water in new concrete that we can tolerate when we want to pour epoxy grout or apply an epoxy coating?" Most of those considered to be experts in the fields of concrete and epoxy grout or epoxy coatings are too wise to pin themselves down by making a flat statement about the percentage of moisture in new concrete.

Where epoxy coatings are concerned, the industry considers 2% moisture in concrete to be a dangerous level. Applying an epoxy coating to concrete with a 2% or greater moisture will affect the curing of the concrete. When the moisture tries to escape from the concrete and meets the resistance of the coating, it can literally blow the coating off by peeling it or bringing the concrete surface off with it. Old concrete is a different matter. Epoxy grout is not much affected by moisture after it is mixed. As long as the old concrete is not saturated with water, the epoxy will develop a good bond to the concrete. However, you still have to think about how the moisture will escape.

The easiest method to determine the readiness of a concrete foundation to have epoxy grout poured on it is to find out how much moisture would collect at the grout to concrete interface, or the bond line. This is easily done by taping a 3 × 3 ft polyethylene

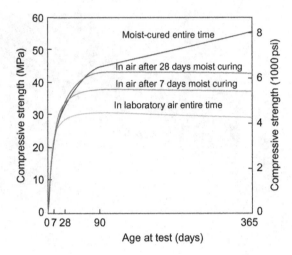

Figure 1.13 Compressive strength gains when concrete kept moist during cure.

sheet to the concrete surface. If moisture collects on the underside of the polyethylene sheet before the epoxy would have time to cure (harden), allow the concrete more time to dry (cure or hydrate) to prevent the possibility of moisture forming at the concrete to grout interface and detracting from the bond of the epoxy to the concrete.

You can perform a shrinkage test as per ASTM C 157 to determine when concrete shrinkage is minimal. If you do not perform a test, you can approximate cure time as follows:

Standard concrete	5-bag mix	21–28 days minimum
High early concrete	6–7 bag mix	7–14 days minimum

Figure 1.13 shows how moist-cured test samples of concrete reached a better compressive strength than air-dried concrete. Samples that remained in a moist condition continued to gain strength.

Conclusion on Concrete Foundations

- Proper curing of concrete is essential to dynamically loaded machinery foundations.
- Epoxy grout will develop a bond to new concrete in excess of the concrete tensile strength.
- The higher the level of moisture in the concrete, the lower the concrete tensile strength.
- The lower the concrete tensile strength, the greater the possibility of a concrete failure just below the epoxy grout to concrete bond line.
- If epoxy grout is poured too early and the moisture is still trying to leave the new concrete after the grout has cured, a delamination may occur just below the grout to concrete interface.
- When concrete shrinkage is minimal, epoxy grout can safely be poured on new concrete.
- You can never wait too long for concrete to cure.

2 Anchor Bolts

The grout holds the machine up, but the anchor bolts hold it down.

Why do anchor bolts fail? This question seems to come up time and time again. There are several reasons for anchor bolt failure.

The anchor bolt may have been improperly designed, installed, was the wrong metallurgy for the service it was in, had insufficient embedment depth, or was misaligned. The anchor bolt could have been improperly maintained. Anchor bolts subjected to severe thermal, dynamic, and cyclic loads, and having unbalanced forces acting upon them should be checked periodically for proper tension. Or, there was inadequate vertical reinforcement of the concrete around the anchor bolt specified during the engineering or design phase of the installation.

To keep bolted connections tight, we can choose from several methods. The most common is the insertion of a locking device between the rotating part (nut) and the parts being fastened. That locking device usually is a split-ring lockwasher. A split-ring lockwasher does not always meet all locking requirements.

The Grouting Handbook. DOI: http://dx.doi.org/10.1016/B978-0-12-416585-4.00002-X

For an anchor bolt to do its job, the following standards must be met:

- The bolt must be designed properly;
- The bolt must be sized properly;
- The bolt must have sufficient embedment depth (10–40 times the bolt diameter);
- The bolt must have sufficient free length for proper tensioning (Minimum of 12 times the bolt diameter);
- The bolt must be kept tight; and
- The bolt must have sufficient vertical reinforcing steel surrounding it in the concrete.

There is a lot taken for granted when it comes to anchor bolts. For years anchor bolt installation has been haphazard at best unless a qualified specialist was at the job site. Many anchor bolt installations have the bolt grouted in with only minimal, or no free length Figure 2.1. To properly tension an anchor bolt, there must be enough unbound length to allow the bolt to stretch. The rule of thumb for this is *"12 times the bolt diameter."*

Most anchor bolt tensioning is usually only as good as the 1¼ in. wrench with a 3 ft cheater pipe that is used. Overtightening (stressing) and lack of proper maintenance are the major causes of anchor bolt problems.

Unfortunately, anchor bolts do break or fail from time to time. The factors that cause anchor bolt failure can be traced back to the initial installation:

- Grout sleeves filled with grout, no available free length (Figure 2.2).
- Grout sleeves not filled with anything except water (or worse) and corrosion takes place over time Figure 2.3.

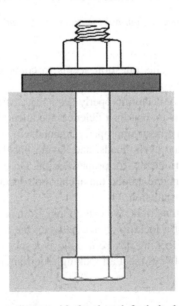

Figure 2.1 No anchor bolt wrap to provide free length for bolt elongation.

Figure 2.2 Bolt sleeve filled with grout.

Figure 2.3 Severe bolt corrosion due to sleeves not being filled with a suitable filler.

Grout sleeves and anchor bolts do not line up with equipment bolt holes. (This one is usually corrected with a nonstandard repair such as a come-along, a torch, or creative thinking as shown in Figure 2.4.

Using nuts on the underside of the machine base to act as the leveling device is not a good idea. These are always left in place when the grout is poured. As shown in Figure 2.5, this type of installation allows for a very small amount of free length for tensioning the anchor bolt.

There have been many advances made in anchor bolt technology. Probably the best has been the use of the *"spherically seated nut and washer"* Figure 2.6. This

Figure 2.4 Noncustomer approved correction for a misaligned anchor bolt by the installing contractor.

Figure 2.5 Nuts used to level a baseplate is not a good idea.

device compensates for slight misalignment of the anchor bolt in its perpendicular plane, and allows for full contact of the nut and washer against the bolted surface. It is a two-piece hardened washer with one side concave and the other side convex that allows it to compensate for up to 7° of anchor bolt misalignment.

Torquing the anchor bolt is preloading it. The length of the bolt controls the load stain curve. The load on a longer bolt is less sensitive to changes in stain. Strain changes could

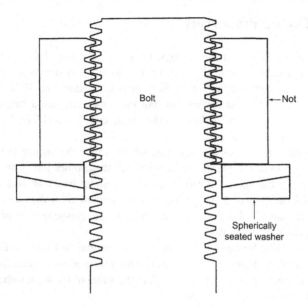

Figure 2.6 A spherically seated washer helps correct for slight anchor bolt misalignment.

be caused by slip of the anchor or shrink in the grout. Anchor bolts are never intended to operate where a gap opens in the clamped joint. That means the load must always stay under the preload. A longer bolt can more reliably maintain a constant preload.

There are several ways to apply the correct amount of tension to an anchor bolt. These range from elaborate load-monitoring devices that consist of a rotating cap that binds when the bolt elongates a predetermined amount as it reaches its design load, to sophisticated hydraulic and mechanical bolt tensioning devices.

Remember the following about anchor bolts:

- Anchor bolts must have elastic properties;
- The shear strength of an anchor bolt is ½ its tensile strength;
- Apply enough preload (clamping force) to overcome an externally applied cyclic load;
- Preload is usually ⅔ of the tensile strength of the bolt;
- Allow ⅓ of the bolt diameter projection beyond the nut on both ends for thread deformation;
- Eighty-five percent of the permanent set in an anchor bolt occurs in the first load reversal; and
- For reciprocating machines the clamping force of the anchor bolt should be at least three times the deadweight of the machine. It is the clamping force and friction that prevents lateral movement (vibration), not grout adhesion.

EXAMPLE: Deadweight 125,000 lb divided by the number of anchor bolts (24) = 5,208 lb times 3 = 15,624 lb of clamping force (or more) per anchor bolt.

What Are Elastic Properties?

The easiest analogy is to remember back to a toy you may have had as a child. Remember the toy called a Slinky? Lay it on the ground and stretch it out so that you have 1 pound of tension (preload) in it. Now pull it an additional 10 in. No big deal, right? Now, let's do the same experiment, but this time only use a single coil of the slinky. Put a pound of tension (preload) in the single coil. No problem. Now, pull it an additional 10 in. Complete failure.

So remembering the Slinky analogy, you simply make the anchor bolt longer and don't bother trying to determine the exact loads that these other forces apply. You simply plan for the fact that there are certain conditions under which an additional (small) displacement will be applied to the fastener and you plan to allow the member to stretch that additional amount. In doing so, you simply make the elastic section of the member longer and let everything work itself out.

For some loads like heating/cooling cycles, vibrational and impact, the load is so large that the strength of the fastener is irrelevant. The load will move the end of the fastener. In these cases, you would not calculate the value of the load; rather you would only consider the resulting displacement.

Measuring the Equipment Pull-Down at the Anchor Bolts

When attempting to measure the amount of equipment pull-down, it is a common practice to use a dial indicator to observe the pull-down effect on a piece of machinery when the anchor bolts are being tightened. The value and significance of the readings depends on how, and where, the dial indicators are positioned. The dial indicator will not give absolute information. It will show differences and relative movements. A typical arrangement will measure the compression of the underlying epoxy grout, and concrete, down as far as the bolt's anchor point. With an anchor bolt having 12 times the bolt diameter of available free length this compression could typically be as high as 0.007 of an inch. The greater the free length of the anchor bolt, the greater the magnitude of the dial indicator reading for any given bolt stress. The relationship is not linear because the stress is progressively distributed within a conical envelope I will call the load cone.

It is apparent why large indicator readings can occur. If the indicator head is positioned to contact the grout at the edge of the chock then only the chock compression will be measured, typically 0.0005–0.0015 of an inch. This compression is not a problem if all chocks have the same amount of pull-down, and the amount of pull-down is less than 0.010 in. between anchor bolts. In the case of a gas engine compressor the alignment (web deflection) can be affected by a change in elevation at the anchor bolt of 0.010 in.. For every 0.010 in. elevation change at the anchor bolt the web deflection of a crankshaft increases by approximately 0.001 in.

If the chock is poured on a steel soleplate or rail then the indicator should be set to read off the steel surface. A soleplate or rail is not significantly deflected by the anchor bolt stress, so only the chock compression will be measured.

As long as all the anchor bolts are of similar length and material, and they are all loaded (or tightened) equally then a machine such as a gas engine or compressor, will not have its alignment adversely affected.

WARNING: Failure to maintain proper anchor bolt tension could result in the following:

- The machine becoming loose.
- The anchor bolt may be placed in a shear condition.
- Depending on the length of the equipment and rigidity of the frame, flexing on a horizontal plane could occur.

Determination of Anchor Bolt Pullout Strengths

Epoxy grouts have been used for many years to secure anchor bolts, dowels and reinforcing bars. This topic is a fairly complex one, and should be handled on a case-by-case basis. The various forces acting on the anchor bolts must be thoroughly understood by the engineer, since he is responsible for the installation. This chapter will address the bolts and anchoring system used to secure machinery to concrete foundations.

One method of approximating the pullout strength of a grouted anchor is to assume that the anchor is secured to the foundation by a cylinder of epoxy grout. The resisting strength is provided by the bond strength of the grout to the concrete across the entire vertical surface of the grout. This method of approximation provides a usable value for embedment depths of less than 8 in.

The pullout load strength can be approximated from: A better approximation can be determined from information presented in the *Guide to the Design of Anchor Bolts and Other Steel Embedments* by R.W. Cannon, D.A. Godfrey, and F.L. Morehead; Concrete International/July 1981, p. 28. If a hex-headed bolt is grouted into a hole and tested, then failure will occur as a *cone* of concrete extracted from the foundation. The pullout load is equal to the bond stresses of the concrete over the surface area of a 45° fracture cone. The average bond strength (fb) of the concrete foundation is obtained from an approximation similar to the tensile strength of the concrete. This bond strength is multiplied by the surface area of the fracture cone to give an approximate value for the pullout load, (P).

P = Average Bond Strength of Concrete $\times$ Surface Area of the Pullout Cone

$P = 2.6 \times (\text{compressive strength})^{\frac{1}{2}} \times 4.443 \times d^2$

The average bond strength of concrete is:

$f_b = 4 \times 0.65 \times (\text{compressive strength})^{\frac{1}{2}}$

$f_b = 2.6 \times (\text{compressive strength})^{\frac{1}{2}}$

The coefficient 4 represents a factor to average the concrete bond stress over the face of the cone. The coefficient 0.65 adjusts the figure by assuring the bolt is near the top or middle of the foundation slab. (The pullout strength varies with how close the bolt head is to the bottom or top of the slab, in addition to several other factors).

The term (compressive strength)$^{1/2}$ means the same as square root of compressive strength.

Therefore, the average bond strength for 4,000 psi concrete is

$$f_b = 2.6 \times (4,000)^{1/2}$$
$$= 2.6 \times 63$$
$$= 164 \, \text{psi}$$

This value is slightly different from the tensile strength because the load is applied as a fractured cone. The tensile forces on the concrete are averaged across the surface of the cone. The stress is greatest at the bolt head and least at the concrete surface, where the cone has its widest expanse.

The surface area of the cone is derived as follows:

Area for a cone is $= \pi r \times (r^2 + d^2)^{1/2}$

Where d represents the depth of embedment, and r represents the radius of the fracture cone.

At a 45° angle $r = d$ area $= \pi d \times (d^2 + d^2)^{1/2}$
$$= \pi d \, (2d^2)^{1/2}$$
$$= \pi d^2 \times 2^{1/2}$$

Substituting $2^{1/2} = 1.414$ $\pi = 3.1416$

Area $= 3.1416 \times d^2 \times 1.1414$

Then the area of the 45° cone is $= 4.443 \times d^2$

The following tables have been developed directly from this formula. The assumptions pertaining to these tables are:

- The compressive and tensile strength of the anchor grout is greater than that of concrete.
- Failure occurs in the concrete substrate, not in the grout or the anchor bolt.
- The shear bond strength of the anchor bolt to the grout, and the grout to the concrete is greater than the tensile strength of the concrete itself.
- A safety factor of 25% of the ultimate load has been used.

NOTE: The hole diameter does not influence the pullout strength of the concrete.

The ultimate tensile load is the maximum achievable load resulting in concrete substrate failure. The concrete fails in tension. This figure assumes no yield or failure in the steel embedment. Proper anchor designing will of course require consideration of the grade and diameter of the bolt.

When epoxy grouts are used the heat deflection temperature (HDT) of the epoxy, operational temperature, and duration of the load must be considered.

Concrete Compressive Strength: 3,000 psi (20.7 MPa)

Embedment Depth		Ultimate Tensile Load		Recommended Pullout Load	
Inches	cm	Pounds	Newtons	pounds	Newtons
4	10.2	10,090	44,800	2,500	11,200
6	15.2	22,700	100,900	5,600	25,200
8	20.3	40,370	179,500	10,000	44,800
10	25.4	63,090	280,600	15,700	70,100
12	30.5	90,840	404,000	22,700	101,000

Concrete Compressive Strength: 4,000 psi (27.6 MPa)

Embedment Depth		Ultimate Tensile Load		Recommended Pullout Load	
Inches	cm	Pounds	Newtons	Pounds	Newtons
4	10.2	11,650	51,800	2,900	12,900
6	15.2	26,220	116,600	6,500	29,100
8	20.3	46,620	207,300	11,600	51,800
10	25.4	72,860	324,000	18,200	81,000
12	30.5	104,520	464,900	26,100	116,200

Concrete Compressive Strength: 5,000 psi (34.5 MPa)

Embedment Depth		Ultimate Tensile Load		Recommended Pullout Load	
Inches	cm	Pounds	Newtons	Pounds	Newtons
4	10.2	13,080	58,100	3,200	14,500
6	15.2	29,420	130,800	7,300	32,700
8	20.3	52,320	232,700	13,000	58,100
10	25.4	81,750	363,600	20,400	90,900
12	30.5	117,720	523,600	29,400	130,900

NOTE: The recommended pullout load is 25% of the ultimate tensile load. You must determine whether this safety factor is appropriate.

Machinery Reliability Starts with the Anchor Bolts

The foundation and grouting system of a machine should be designed to resist the downward forces of the equipment deadweight and the forces generated by the tensioning of the anchor bolts. Anchor bolts, on the other hand, are designed to resist the upward and lateral forces produced by an operating piece of machinery. More simply stated *"The grout holds the machine up while the anchor bolts hold the machine down."* Problems with correct anchor bolt tension have existed for years, and will greatly affect the ability of an anchor bolt to do its job.

The majority of the reciprocating equipment used today is installed on some type of chocking system, be it steel, composite, or epoxy. The coefficient of friction (COF) of the composite and epoxy chock provides a greater measure of lateral resistance than does a steel chock. However, the soundness of all chock support systems depends mainly on correct anchor bolt tension and residual preload.

How Anchor Bolts Behave

Large reciprocating engines and compressors typically represent the most severe service for anchor bolts. We will cover some of the problems encountered with, but not limited to, anchor bolts commonly used in these machines.

How many times have we heard, or been involved in the following conversation?

"Did you tighten the anchor bolts?"

"Yeah."

"What torque did you give 'em?"

"What the book called for."

"Are you sure they're tight?"

"Yeah I'm sure ... I torqued 'em."

Sound familiar? But, is what we are trying to achieve simply anchor bolt tightness or are we trying to introduce into the anchor bolt a residual force by stretching the anchor bolt to provide that residual force?

Let's start off by looking at some common misconceptions about bolting in general:

1. A torqued anchor bolt will never loosen.
   Wrong: When initial preload is lost and friction in the threads and under the nut face starts to drop, it doesn't take long for a bolt to loosen
2. It takes thousands of hours of vibration to loosen a bolt.
   Wrong: After side sliding (transverse slipping) starts, as few as 100 cycles may loosen a bolt.
3. I know it's tightened because I torqued it.
   Wrong: 80% of the torque effort is expended to overcome friction, and not to achieve tightness.

4. High torque loads automatically mean high clamp loads.
   Wrong: *85% of the tightening torque is absorbed in the threads and under the nut. Only 15% produces clamp load.*
5. Split-ring lock washers exert an auxiliary pressure on the underside of the nut face that prevents loosening.
   Wrong: *They have no effect.*
6. Double nutting will keep the applied tension on the bolt.
   Wrong: It has no effect. (See Figure 2.5)

Definition of Terms

The first thing we need to do is to get an accurate definition of the terms we will be using. Webster's dictionary gives several definitions for torque and several for tension. The ones that relate to what we will be discussing say the following:

Torque is:
A twisting or wrenching effect or moment exerted by a force acting at a distance on a body, equal to the force multiplied by the perpendicular distance between the line of action of the force and the center of rotation at which it is exerted.

Tension is:
Stress on a material produced by the pull of forces tending to cause extension. (In the case of an anchor bolt this will be its stretch or elongation from a relaxed state.)

Obviously, there is a big difference between torque and tension, torque being the act and tension being the result. The next set of terms we should define will be initial preload and residual preload. Webster's dictionary gives about 20 terms, more or less, for load. None of these terms seem to apply to anchor bolts so I will attempt to define them as follows:

Initial preload: The clamping force imposed on the machine by the anchor bolt when it is first tightened, from a relaxed state.
Residual preload: The clamping force remaining on the machinery after external forces or conditions have affected the initial preload over time. The residual preload by definition will always be less than initial preload.

Defining the above two terms introduced a new term to define:

Clamping force: The amount of compression between the underside of the anchor bolt nut and the anchor bolt embedment point in the foundation.

The clamping force of an anchor bolt is dependent on the residual preload existing in that bolt at that particular time, not the initial preload that was originally introduced to the bolt.

Factors Affecting Preload

Anchor bolt preload can be increased or decreased by various factors. The mechanisms that can cause these changes from initial to residual preload, are common, but

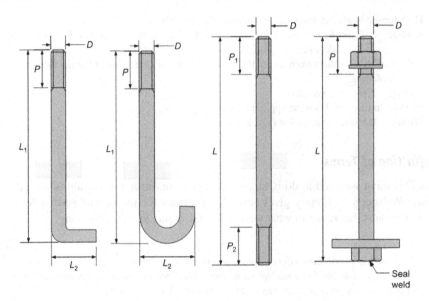

Figure 2.7 Common anchor bolt installation arrangements.

frequently go unnoticed. Getting the correct preload or tension the first time around can be difficult to achieve. If the mechanic is using torque to achieve the tightness of the bolt, there can be factors of which he may not be aware that will affect what he is trying to do ... applying torque, and what the ultimate goal is ... getting the proper preload or tension in the bolt.

The following are the most important of the many factors affecting torque as a way to achieve proper anchor bolt tensioning.

The anchor bolt and its accessories:
- Hardness of all anchor bolt components. (The nut should be harder than the washer to prevent galling.)
- The anchor bolt material. (Correct material for the ultimate load or stress applied.)
- The size of the anchor bolt.

Installing the anchor bolt: (Figure 2.7)
- Perpendicularity of the anchor bolt to the surface being clamped. (The force from the nut should bear equally on the machine frame for 360°.)
- The amount of available free stretch length for the anchor bolt. This is described as the area of the bolt that is not encapsulated by the concrete. (Usually a minimum of 12 times the bolt diameter. Actually the more available bolt length for stretch the better.)
- Bolt embedment depth into the foundation. This is the amount of anchor bolt length that is encapsulated by the concrete. (Usually a minimum of 10 times the bolt diameter.)

Lubricating the anchor bolt prior to tensioning:
- Type of thread lubricant used to reduce friction between seating surfaces and between the threads. See Table 2.1

Factors affecting torquing of the anchor bolt:
- The competency of the person doing the torquing;
- The condition of the torque wrench;

Table 2.1 This Table Gives the COFs of Various Types of Thread Lubricants

Lubricant	COF	Percentage Effort Lost to Friction	Percentage Effort to Tension Applied	Relative Torque in ft lb for a 1 in. Bolt at 50% Minimum Yield
Moly/oil	0.060	83.1	16.9	642
Lead/oil	0.09	88.6	11.4	945
Copper and graphite with oil	0.100	88.6	10.8	998
Steel on steel dry	0.400	97.05	2.95	3,669

- Temperature at which the bolt is torqued (When an anchor bolt is initially tightened, it is done when the piece of equipment is shut down and it is cool enough for someone to safely work on it. In the case of an overhaul, or new installation, the entire unit is at a "cold iron" condition.); and
- The ultimate temperature at which the bolt will operate.

Once the equipment is started up and placed on line, the anchor bolt can expand 0.018 in. or more as it goes from its initial temperature, which in some cases could be as low as 30°F, up to operating temperature which normally is around 155–165°F. This adds up to a thermal differential of around 125–135°F. The amount of growth experienced by the anchor bolt will depend on the coefficient of thermal expansion of the bolt material. It is possible for the anchor bolt to actually become loose simply by thermal growth. Very seldom, if ever, is the equipment shut down and the alignment or anchor bolt tension rechecked.

Under field conditions, most of these factors are not even considered, much less corrected or controlled. These days, it is the perceived economics of the situation that control what is to be done. Unfortunately, the economic perception will not allow "enough" control to guarantee the results we would like … correct and lasting anchor bolt tension. Many of the factors mentioned above can cause a loosening of the bolt, or nut, to start within 24 h after the initial preload or tensioning is applied. In many situations, the loss of preload or tension is ignored, not because the mechanic didn't do his job, but because there is no accurate way in which to monitor and measure tension gain or loss after the initial preload was introduced.

It can thus be stated that anchor bolt problems are largely related to:

- Achieving the correct initial preload, or tension, in an anchor bolt.
- Maintaining the desired preload, or tension, over time.
- Accurately monitoring the desired preload or tension of the anchor bolt.

With the technology available today we can measure and observe the various factors that cause preload loss after initial tightening. However, unless there is a routine anchor bolt maintenance program in effect at your facility, it doesn't matter what device or method you use to obtain the initial preload. If the bolt is not initially tensioned properly and checked at least twice a year for residual preload, you cannot be sure the bolt is doing its job.

Conventional Procedures for Achieving Initial Anchor Bolt Tension

- Tension and release the bolt two times. Perform final tensioning or preload on the third try.

> **NOTE:** The amount of time between these tensioning's will depend on the anchor bolt material its elastic properly, and temperature. This should range from a few minutes to many hours for the bolt to relax from the stretch imposed on it.

- Check the anchor bolt for proper tension and make any necessary adjustments 7 days after the equipment has been placed in service.

> **NOTE:** The anchor bolt is not loosened for this or any other tension checks unless it is grossly overtensioned. If the bolt is overtensioned, loosen it only enough to return it to the required or designed preload. Never completely loosen and then retension an anchor bolt unless it is at ambient temperature. This will ensure that the bolt will be at the design preload at its operating temperature.

- Thirty days after the initial tensioning, recheck the anchor bolt for proper tension and make adjustments with the equipment at operating temperature.
- Six months after initial tensioning, check the anchor bolts for proper tension and make adjustments.
- Check the anchor bolts for proper tension every 6 months thereafter and make adjustments.

The conventional methods are rather complicated and not very practical given the amount of manpower and time required. Also, pull-down or rise of the machine at each anchor bolt location should be monitored each time the bolts are checked for proper tension. Record and plot these readings. If you note excessive pull-down, recheck the machine alignment.

Types of Material Normally Used for Anchor Bolts of 1–1½ in. in Diameter

- Mild steel, ASTM A36 (Grade 5) bolt material, minimum tensile strength 105,000 psi;
- High-strength, ASTM A193 (Grade 7) bolt material, minimum tensile strength 133,000 psi. (This is the material commonly used in today's high-strength anchor bolt systems.);
- Aircraft or nuclear-quality, ASTM A286 (Grade 8) bolt material, minimum tensile strength 150,000 psi; and
- Above Grade 8 bolt material is ultra-high-strength alloy steel, minimum tensile strength 185,000 psi.

Typical Ways of Applying, or Measuring, Anchor Bolt Preload or Tension

Sledgehammer Method

Using a hammer wrench (Figure 2.8) and sledge is a crude and dangerous method used to tighten large anchor bolts of 1½ in. in diameter and greater. This method gives little control over the amount of tension the *bolt* receives, is very inconsistent, and is a common cause of injuries.

Torque

Using torque as a measurement to estimate anchor bolt preload or tension has never been particularly reliable because of some of the factors mentioned previously. Torque measures only the torsional effort applied to the bolt or nut. How much of this effort is transmitted into actual bolt tension depends on the mating interface friction. Friction is unpredictable because of many variables such as lubrication, cleanliness, and thread condition.

Loosening Force

Measurement of the preload or tension after the initial tightening of an anchor bolt has always posed difficulty in obtaining accurate readings.

Some people still try to do this by measuring the amount of force it takes to loosen an anchor bolt. (This is sometimes referred to as breakaway torque.) This method is

Figure 2.8 Hammer wrench.

based on the old school of thought that anchor bolts take more torque to loosen than to tighten. This is not an accurate method, however. Due to variations in static friction, you cannot know exactly when the actual loosening, or breakaway, has occurred. It is not possible to use the visible motion of the nut or the noise it makes when it finally breaks loose. The nut usually "pops" when it breaks loose. But if you use a strain gage, you see that a change in the bolt tension precedes or follows what is seen or heard.

Ultrasonics to Measure Bolt Length

Using ultrasonic technology to measure the bolt length in a relaxed state (Figure 2.9) versus the bolt's elongation during and after tensioning will enable you to know the differential length. In some applications, you can compare this differential length to residual preload. To do this, you must record the bolt measurements to provide a base reference for future measurements. These notes must contain a permanent record of the initial acoustic length of each anchor bolt that is tightened and its location. By numbering and marking each bolt, you will know the one to which you are later returning. If you have kept such a log, you can return to the bolt at any time to

- Remeasure its current length ultrasonically;
- Compare this reading to the initial length before it was tightened; and
- Calculate the stress levels in the fastener regardless of how much time has occurred since initial tightening.

If you have not used ultrasonic technology for the initial tightening of the anchor bolt, you can measure the decrease in the overall length of a few anchor bolts as they are loosened. This decrease may give you some idea of the amount of residual preload that was in the bolt before it was loosened.

Figure 2.9 A technician gathers preinstallation relaxed bolt length.

Seven Possible Causes of Loss of Preload in Anchor Bolts

Anchor Bolt Relaxation

This is a short-term effect that can occur over a relatively short period of time when compared to the life of an anchor bolt. It can account for only a few percent loss in initial preload. About 5–10% is not uncommon, although relaxations as high as 25% have been observed.

When new anchor bolts are first tightened, the thread surfaces make contact only on microscopically small high spots. Because of variations in machining accuracy, these surfaces are never perfectly flat. Anchor bolts and their associated parts are designed to support high surface loads but only if substantial portions of the surfaces are engaged. The high spots on the thread surfaces will always be overloaded, past their yield point, when initially tightened. These high spots will break down (creep), and this process will continue until a large enough percentage of the available contact surfaces has been engaged to stabilize the process.

Retightening the bolt can partially compensate for this type of relaxation. After the parts have worked together, they will no longer creep, or at least not as badly as they do when new.

Thread Deformation

Plastic flow and stress will occur in thread roots and in the first few bolt threads that engage the nut threads. An old rule of thumb was that for every 10,000 lb of stress applied to a bolting mechanism, the load was picked up by an additional thread. An example of this would be a nut containing 10 threads. If it was threaded completely onto the bolt and the assembly tensioned to 30,000 lb, it would be carried by three threads. If the load were being increased by 10,000 lb four threads would now be carrying the 40,000 lb load.

Anchor Bolt Misalignment

If this condition exists, it is highly unlikely that the anchor bolt in question will stay tight, let alone develop the required clamping force. This misalignment (Figure 2.10) can be as little as 1½ a degree off of the vertical to have a detrimental effect on the life of an anchor bolt and its ability to maintain the desired clamping force or tension.

When the misaligned anchor bolt nut is tightened, it places a bending moment on the bolt, which is not the desired tension mode. If the misalignment is severe enough and a high clamping load is placed on the bolt, the stress concentration placed on the thread root of that anchor bolt will eventually cause the ultimate failure of the bolt, and high replacement or repair costs will result.

Anchor bolt misalignment is particularly damaging to anchor bolts with cut threads. Here are the warning signs you should look for:

• Figure 2.11 shows the anchor bolt is not perpendicular to the mating surface it is holding down. If this is more than 7°, you will not be able to easily fix this.

Figure 2.10 Misaligned anchor bolt. This was caused by an improper installation on the initial foundation pour.

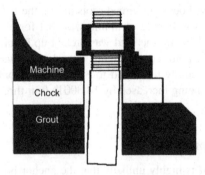

Figure 2.11 Bolt is not perpendicular to the mating surface it is holding down.

- The face of the nut is not exactly perpendicular to the axis of the threads. Almost the same as the preceding point except that this could be a case of poorly machined threads.
- A corner of the nut makes contact with a poorly machined mating surface before the nut face does.

Depending on the material of the nut and machine base, this misalignment condition could also result in galling of the nut or base and the loss of a good, flat surface. Using a two-piece spherically seated washer will just about eliminate the problem of bolt alignment providing the angle does not exceed 7°.

For whatever reason, if the anchor bolt nut face does not bear equally for 360° against the machine mating surface, stress concentration in the thread root, resulting in plastic flow and ultimate bolt failure, will occur as shown in Figure 2.12.

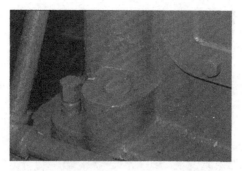

Figure 2.12 Nut only making contact on one side caused rapid failure at the thread root for this anchor bolt.

Figure 2.13 Resin-rich grout surface is caused by removing aggregate to improve flow.

Grout Creep

Grout creep is another cause for short-term relaxation. Most epoxy grouts will deform to some extent when subjected to the following:

- Higher temperatures than those for which the grout was designed.
- Higher loading than that for which the grout was designed. (This can also be an effect of higher temperatures.)
- Reduced aggregate fill ratios to make the grout more flowable during placement. This changes the physical properties and reduces the load carrying capability of the grout. It allows the aggregate to settle to the bottom as shown in and not remain evenly distributed in the grout. Figure 2.13 shows how the aggregate has settled, with the larger aggregate on the bottom and getting progressively smaller toward the top.
- Excessive air is entrained in the grout during initial mixing. Figure 2.14 shows how bad this can be. Overmixing the liquids (foaming) and overmixing the aggregate (continuing to mix after the aggregate has been completely wet) can contribute to this. Air usually finds its way to the surface just prior to the grout hardening.

Grout creep is an often-used excuse for poor anchor bolt maintenance or epoxy grout installation.

Figure 2.14 Air whipped into the grout will find its way to the surface.

Elastic Interactions

In some cases, short-term relaxation may occur between anchor bolts and frame members as you tighten a group of anchor bolts.

When you tighten the first bolt in a series, stretch it and place the immediate vicinity of that bolt in compression. When the next bolt is tightened, compress the area of the first bolt a little more. This additional compression of the frame allows the first bolt to relax a little. Even if you achieved perfect initial preload in these two bolts when you tightened each; you would now find that only the second bolt had the correct preload. Just by tightening that second bolt, you have eliminated some of the initial tension you put into the first bolt.

The amount of elastic interaction that could be experienced in a frame or structural support will depend upon such things as the stiffness of the frame and its members, the size of the bolt flange, the distance between bolts, and so on. Elastic interactions can create substantial losses of initial preload. Losses of 40–60% have been observed.

You can compensate for elastic interaction in three ways:

1. Apply more tension to the first bolts tightened than to the last ones. The act of tightening the last draws down the first but doesn't eliminate all preload. Using this method will usually result in some bolts being tightened too much. This in itself can be a problem because in applying more tension to the first bolt than to subsequent ones, you can wind up taking a bolt past its yield point. In most cases, you are trying to tighten the anchor bolts to some point near their yield strength.
2. Apply a modest amount of torque to the "low" bolts in the final step of a procedure requiring several passes. Even this can result in overtightening of some of the anchor bolts. If that final torque is proper, it will not drive all of the preload out of the "high" bolts.
3. Use ultrasonics to measure the elongation/preload after all bolts have been tightened. Apply necessary tension to correct those bolts that have lost any preload.

Like embedment relaxation, elastic interactions are very common but unnoticed. Maintaining uniform residual preload in an anchor bolt system along a frame is difficulty if you are trying to apply initial torque or tension to those bolts, in a pattern or sequence, over several tries.

Vibration Loosening

Vibration loosening can eliminate all initial preload in an anchor bolt.

Under severe vibration, the bolt at first loses preload slowly over a period of time. After the residual preload in the bolt drops below a certain point, it can no longer prevent side sliding (transverse slippage) between the anchor bolt nut and frame flange surfaces.

Anchor bolts can lose 20–40% of initial preload when subjected to vibrations that are perpendicular to their axes: This transverse vibration (side sliding) is a real problem and can cause complete loss of initial bolt preload.

You can prevent vibration loosening, but only if the anchor bolt system is properly designed and installed. Select a properly designed bolt that can develop sufficient preload to resist transverse slippage by taking the following design elements into consideration:

- Rolled threads
- Heat-treated materials
- Spherically seated washers
- Hardened nuts
- Maintaining the proper bolt preload.

Stress Relaxation

This is a long-term event and becomes a problem only in applications where the anchor bolts are subjected to extreme temperatures. High-temperature differential can cause the atoms in an anchor bolt to realign themselves over time to reduce the high-tensile stresses they are placed under during initial bolt tensioning.

Stress relaxation is just a different form of relaxation you know as creep.

Creep occurs when a material is subjected to a high temperature and a high constant load. Over time, the bolt material will gradually creep (change dimension) because of atom realignment.

Anchor bolts must be given high loads to do their job. When the bolt is subjected to high temperature under these high loads, it will gradually lose tension or preload without any physical change in its length. The only way you can avoid stress relaxation is to use anchor bolt materials that have a high resistance. Aircraft engine and nuclear designers favor materials such as A286 for high-temperature applications.

Bolt Tensioning Devices Currently Available

Multi-Jackbolt Tensioners (MJTs, sometimes referred to as jackscrew tensioners or Superbolts®).

These are used in lieu of a conventional nut (see Figure 2.15). MJTs require only "hand tools" to install or remove. Rather than turn an entire hex nut or bolt with high-powered tooling, MJTs are installed by turning the individual jackbolts. Each jackbolt "pushes" against a hardened washer, and the opposite reaction force of the main bolt head creates a strong clamping force on the flange. Because the jackbolts have a small friction diameter, a high thrust force is created with relatively little torque input.

Figure 2.15 MJTs, sometimes referred to as jackscrew tensioners or Superbolts®).

A Comparison Between MJTs and Common Hex Nuts

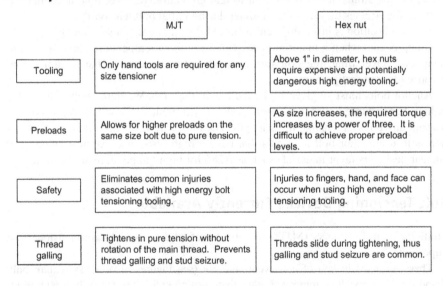

	MJT	Hex nut
Tooling	Only hand tools are required for any size tensioner	Above 1" in diameter, hex nuts require expensive and potentially dangerous high energy tooling.
Preloads	Allows for higher preloads on the same size bolt due to pure tension.	As size increases, the required torque increases by a power of three. It is difficult to achieve proper preload levels.
Safety	Eliminates common injuries associated with high energy bolt tensioning tooling.	Injuries to fingers, hand, and face can occur when using high energy bolt tensioning tooling.
Thread galling	Tightens in pure tension without rotation of the main thread. Prevents thread galling and stud seizure.	Threads slide during tightening, thus galling and stud seizure are common.

Hydraulic Tensioning Devices

This system has a relatively easy setup. The hydraulic pump that provides pressure to the head can be either air or electric (Figure 2.16). It does develop a somewhat

Figure 2.16 Hydraulic tensioner.

uniform clamping force without the torsional effect, but you can easily overstretch an anchor bolt and cause permanent bolt deformation unless your calculations and the pressure gage are accurate. You can only know what the bolt tension is by presetting the hydraulic (stall) pressure on the pump unit during the initial setup phase of the operation. When you transfer the load from the hydraulic jacking device to the mechanics of the nut and bolt, relaxation (over time) will occur, and retensioning may be necessary. This load transfer relaxation is caused by equipment heating and/ or cooling and is unpredictable and variable.

AUTHOR'S NOTE: I have personally used hydraulic tensioners and have had a few close calls with them. The user should be aware that they are only ½ second away from severely injuring or losing a finger with this type of tensioning device. Visit this web site for safety instructions for hydraulic tensioners.
  http://bolting911.com/media/pdf/Safety-Instructions.PDF

Torque Multipliers

Torque multipliers are generally a stacked system of multiple ratchet-type wrenches with the final piece being a torque wrench (Figure 2.17). The ultimate reading on the torque wrench is then multiplied by a factor. These tools are space-limited, require a reaction member, are bulky, and it is difficult to maintain force perpendicularity while using them. This type of tool is commonly used by those not understanding the dynamics of tensioning and often gives the user a false sense of accomplishment.

The torque multiplier shown in Figure 2.18 has a ½ in. female input and a 1 in. male output. Maximum input 173 ft lb will result in a 3,200 ft lb output. This unit has an accuracy of ±5%. The approximate cost of a torque multiplier is $1,000 per

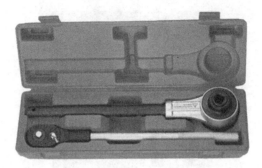

Figure 2.17 Typical torque multiplier.

Figure 2.18 High-output torque multiplier.

1,000 lb of torque generating capability. These are only as good as the calibration of the torque wrench.

NOTE: None of the preceding devices by themselves will indicate if the bolt is overtensioned.

The Easy Way to Observe and Monitor Anchor Bolt Preload, Both Initial and Residual

The best method of obtaining and monitoring proper anchor bolt preload or tensioning is by its stretch. You can monitor the stretch and ultimate load or clamping force exerted by the bolt in several ways. Some of the methods we have discussed are

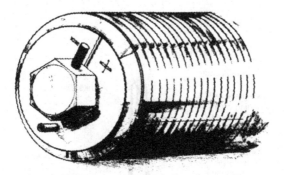

Figure 2.19 MagBolt with load indicator.

Figure 2.20 Rotabolt with a spin-type indicator.

quite detailed and require extensive record keeping. Moreover, they tend to be labor-intensive and time-consuming. Some anchor bolts on the market are equipped with a permanent indicator to let the mechanic know when the correct preload is obtained. These bolts require some preengineering in the form of determining exactly what preload is required. The end user must tell the manufacturer what preload will be required so the indicator can be calibrated to show when the bolt reaches that point. Two basic types of load-indicating anchor bolts exist: the MagBolt® (Figure 2.19) and bolts equipped with the Rotabolt® device (Figure 2.20).

These Two Systems Have Several Characteristics in Common

- Both of these devices mechanically monitor the bolt elongation during tensioning like a micrometer or dial indicator.
- Any bolt receiving these devices must be drilled and tapped in its center on the nut (top) end for positive anchoring of the device at a depth within the bolt to allow for accurate stretch measurement.
- Both of these devices are precalibrated for the ultimate load or tension required.
- During installation, it is *critical* that these bolts be installed at an elevation that will allow only a few threads (three to four) to protrude above the top of the nut. Each ¼ in. of bolt projection above the nut will decrease the accuracy by approximately 5%.
- Both of these assemblies can indicate whether the bolt is overtensioned or undertensioned.

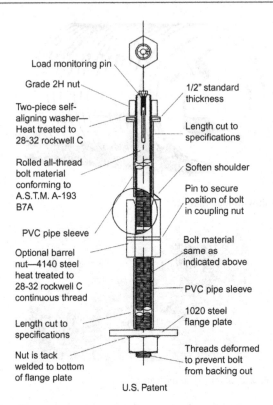

Load monitoring pin

Grade 2H nut

Two-piece self-
aligning washer—
Heat treated to
28-32 rockwell C

Rolled all-thread
bolt material
conforming to
A.S.T.M. A-193
B7A

PVC pipe sleeve

Optional barrel
nut—4140 steel
heat treated to
28-32 rockwell C
continuous thread

Length cut to
specifications

Nut is tack
welded to bottom
of flange plate

1/2" standard
thickness

Length cut to
specifications

Soften shoulder

Pin to secure
position of bolt
in coupling nut

Bolt material
same as
indicated above

PVC pipe sleeve

1020 steel
flange plate

Threads deformed
to prevent bolt
from backing out

U.S. Patent

Figure 2.21 MagBolt® assembly.

Rotabolt®: Tensioning a bolt with this device does not require a torque-measuring instrument. Check the load by simply twisting the inner or outer "control caps" with your fingers. If the outer cap turns, the bolt has exceeded its upper design limit. If the inner cap turns, the bolt has lost its minimum designed preload.

MagBolt®: This device does not require a torque-measuring instrument to check its load. It uses a finger-operated indicator; however, this device can have one or several hash marks permanently stamped into the head of the bolt. The hash mark is installed at the factory during calibration when the bolt is hydraulically stretched to the desired tension. This unit can be purchased as a complete assembly as shown in (Figure 2.21) or as individual components. The two-piece, upper and lower sections are rolled-thread 4140 A193 B7 all-thread bolt with connecting coupling nut. This will allow for partial replacement and can be used with broken existing anchor bolts.

As shown in Figure 2.19, if the pin points toward the negative symbol (−) to the left of the hash mark, the bolt's tension is lower than design levels. If the pin points toward the positive symbol (+) to the right of the hash mark, the bolt's tension is higher than design levels.

> **NOTE:** The MagBolt® will allow the user to know how much tension above the designed preload the bolt has on it and allows for the installation of multiple hash marks for different loads. The cost of this unit is approximately $150.

> **COST:** The cost differential between the MagBolt® and the Rotabolt® appears to be, on average, $60, with the MagBolt being the most cost-effective.

Load-Generating Washers

Load-generating washers (sometimes referred to as Belleville washers or Belleville springs) have been used on all types of machines since the 1700s. Certain industries use them widely, whereas other industries that could benefit from this technology reject it. A load-generating washer is a cone-shaped disk (Figure 2.22) that will flatten or deflect at a given rate. The deflection rate is usually very high, allowing the washer to produce and maintain very large loads in a very small space.

Load-generating washers are used in a variety of applications where high spring loads are required. They are particularly useful to solve problems of vibration, differential thermal expansion, and bolt relaxation.

Because of the conical design of these washers, a positive constant force is exerted against the anchor bolt nut. Combinations of these washers can allow the end user to increase or decrease the load generated by them on the anchor bolt (Figure 2.23).

Figures 2.22 and 2.23 show a pronounced dish. However, most Bellevilles actually have shallow dishes. Sometimes, it's hard to tell which end is up. If that happens to you, lay the washer on a flat surface and look at it from the side.

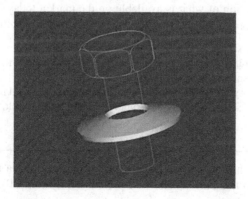

Figure 2.22 Belleville load-generating washer.

Figure 2.23 Stacking Belleville washers allow you to increase or decrease the load applied.

Attributes of Load-Generating Washers

- This force is not affected by the normal temperatures encountered in reciprocating engines or compressors.
- Load-generating washers will not creep.
- Load-generating washers will not experience relaxation.
- Load-generating washers can indicate when a bolt is not tensioned enough.

Ultimate Systems

Based on the discussions earlier in this chapter, the most cost-effective and reliable system is a combination of what is available on the market today. The following are the most important requirements for an adequately engineered bolting system:

- Using a properly designed and installed anchor bolt;
- Using an anchor bolt with an installed load-indicating system that will allow positive, negative, or multiple load readings;
- Using a tensioning system that is easily applied and requires standard hand tools; and
- Using spherically seated washers to compensate for anchor bolt, thread, or surface misalignment.

> **NOTE:** There is not enough data available to support using a load-generating washer system to maintain anchor bolt preload on dynamically loaded anchor bolts of 1 in. and above.

Proper Anchor Bolt Tensioning

The entire reason for tightening an anchor bolt is to provide an additional load that, combined with the equipment deadweight, will hold a piece of equipment in a predetermined position regardless of the designed operating forces that act upon it. When tightening an anchor bolt, most people give little or no thought to what effect their actions will have. If a wrench with a "cheater" extension or an improperly calibrated torque wrench is used to tighten an anchor bolt, the force applied may result in the epoxy chocks being loaded far above or below their design. The condition of the anchor bolt threads and the type of lubricant used can have a dramatic effect on the bolt tightening and resultant chock loads as well. Some older published bolt torques were based on *dry* assembly. Variables such as lubrication, plating, thread form, and so on may increase or decrease applied torque values by as much as 20%, and must be taken into consideration.

Table 2.1 lists the COFs of various lubricants commonly used as thread lubricants. It also shows the amount of effort applied and lost.

A bolt elongates as it is tightened. This elongation can be as much as 0.001 in. of total bolt length for each 30,000 psi of induced tension.

In the case of epoxy chocks, most are designed to carry a minimum load of 500 psi. Improper anchor bolt maintenance can result in a loose chock. When an anchor bolt is initially tightened, it is done when a piece of equipment is shut down and is cool enough for someone to safely work on it. In the case of an overhaul, the entire unit is "cold iron." After the equipment is started and placed on line, the anchor bolt can expand as much as 0.018 in. as it goes from its initial temperature, which in some cases could be as low as 30°F, up to operating temperature, which normally is approximately 155–165°F. This adds up to a thermal range of approximately 125–135°F. This amount of growth will depend on the coefficient of thermal expansion of bolt material.

The anchor bolt could actually come loose simply by thermal growth. Normally when this happens, the epoxy grout or epoxy chock gets the blame for not being able to hold the engine. Epoxy grout or chocks are designed to be in compression; they are not and never will be designed to act as Super Glue to hold down a piece of equipment. Holding the piece of equipment in place is the job of the anchor bolt. Supporting the equipment at a desired elevation is the function of the grout. Very seldom, if ever, is the equipment shut down so the alignment or anchor bolt tension can be rechecked.

The Following Rules Apply for Proper Anchor Bolt Tensioning

- Use the proper type and grade nut for the anchor bolt being used.
- Thoroughly clean the threads, nut face, and flange where the nut face bears.
- If a rough surface is found, dress it out to as smooth a surface as possible.
- Equipment bolt holes and anchor bolts should line up. The distance between the anchor bolt and the vertical face of the bolt hole should be equal on all sides.
- The nut face and washer should bear evenly for 360°. Misalignment of the anchor bolt by as little as 1° off its vertical axis can result in loss of resistance to fatigue.

- Flat washers used between the nut face and the equipment will reduce galling if the washer hardness is less than the nut.
- The lubricant used on the anchor bolt and nut threads must be suitable for the service in which it will be placed. The pressure developed between the metal faces of the threads and washer will range from 25,000 to 50,000 psi. The lubricant must be able to withstand this pressure and not squeeze away or break down. Lubricants containing high percentages of molybdenum disulfide have a bearing pressure limit that allows anchor bolt tightening to a stress equivalent of 100% of yield.

Monitor pull-down at each anchor bolt each time the bolts are checked for proper tension. Record and plot these readings. If you record excessive pull-down, recheck the machine alignment.

The best method of obtaining proper anchor bolt tensioning is to monitor its stretch. You can monitor the stretch and ultimate load or clamping force exerted by the bolt in several ways. One method is an indicator pin mounted in the top of the anchor bolt. Others include load-monitoring washers or mechanical bolt tensioning devices used in lieu of a conventional nut.

Can We Overtighten the Anchor Bolt So We Don't Have to Come Back and Retighten It?

This question doesn't make much sense in light of what has been presented up to now. Up to yield point, the strain in a bolt is proportional to stress. Beyond the elastic limit, the bolt goes into its *plastic range,* causing some permanent stretch to occur. The bolt at this point will not return to its original length; yet the residual tension is fully maintained. It is this residual tension force that keeps a bolt tight and maintains the clamping force on a piece of machinery.

Permanent set in an anchor bolt starts at the section with the highest unit stress. This is usually the thread area below the nut. This thread deformation throws the thread pitch off, which locks the nut and subjects the bolt to tension. This force disappears with wrench removal.

What is the bottom line of all this? A bolt can be tensioned well into its plastic range *IF* it will not be reused or ever need retensioning.

Anchor Bolt Data

Grades

B7	Alloy steel, AISI 4140/4142 quenched and tempered
B8	Class 1 stainless steel, AISI 304, carbide solution treated
B8M	Class 1 stainless steel, AISI 316, carbide solution treated
B8	Class 2 stainless steel, AISI 304, carbide solution treated, strain hardened
B8M	Class 2 stainless steel, AISI 316, carbide solution treated, strain hardened

Mechanical Properties

Grade	Size	Tensile ksi, Min	Yield, ksi, Min	Elong, %, min	RA% min
B7	Up to 2½	125	105	16	50
	2⅝–4	115	95	16	50
	4⅛–7	100	75	18	50
B8 Class 1	All	75	30	30	50
B8M Class 1	All	75	30	30	50
B8 Class 2	Up to ¾	125	100	12	35
	⅞–1	115	80	15	35
	1⅛–1¼	105	65	20	35
	1⅜–1½	100	50	28	45
B8M Class 2	Up to ¾	110	95	15	45
	⅞–1	100	80	20	45
	1⅛–1¼	95	65	25	45
	1⅜–1½	90	50	30	45

Reduction of area (RA)% the difference between the original cross-sectional area of a tensile test specimen and its minimum cross section after the test sample has fractured.

Recommended Nuts and Washers

Bolt Grade	Nuts	Washers
B7	A194 Grade 2H	F436
B8 Class 1	A194 Grade 8	SS304
B8M Class 1	A194 Grade 8M	SS316
B8 Class 2	A194 Grade 8	SS304
B8M Class 2	A194 Grade 8M	SS316

Conclusion on Anchor Bolts

- Correct anchor bolt design and installation plays a critical part in how the preload is obtained and maintained.
- Monitoring its stretch is the best method of obtaining proper anchor bolt tensioning.
- You can obtain and monitor the stretch and ultimate load or clamping force exerted by the bolt in several ways.
- You can use several mechanical bolt tensioning devices to obtain and determine anchor bolt preload.
- Initial preload losses in anchor bolts are due to differential bolt temperatures caused by seasonal and operating temperature changes. You must routinely check the anchor bolts in critically aligned machinery for proper tension or preload.
- A combination of components could provide the user with a system that will indicate both over and under tensioning, compensate for misalignment and maintain adequate preload on the anchor bolt.

3 Cement Grouts

A Comparison Between Epoxy and Cement Grouts, or Which Product Should We Use?

This question comes up time and time again. Epoxy and cement grouts each have preferred usages in today's industrial construction and some cases overlapping as well as specific applications. These are distinctly differing systems, with quite different chemistries, and need to be sold and serviced as their individual application dictates. In most cases, recognizing the differences is the key to success or failure of an installation.

In general, epoxy grouts are recommended for:

- Very high early strength applications resulting in lowest downtime.
- Chemical and oil resistant and in some cases as a corrosion preventer.
- Vibration damping.
- Grouting reciprocating compressors.

The Grouting Handbook. DOI: http://dx.doi.org/10.1016/B978-0-12-416585-4.00003-1

- Applications requiring maximum bond to foundation and machine or baseplate resulting in a monolithic structure.
- Injection into cracks and deep penetrations for repairs to concrete.
- Ability to withstand high dynamic loading without failure.

Cement grouts are recommended for:

- General civil construction, erecting baseplates, grouting precast members or providing bedding mortar, column joints, etc.
- Anchor bolts cables or rod grouting not requiring sustained high tensile loads.
- Applications of machinery or equipment with temperature cycling or sustained elevated temperatures over 200°F.

Applications where both epoxy and Portland cement grouts may be used include:

- Any installation of noncritical rotating equipment or where long-term equipment reliability and operation is not a major concern.
- Structural elements requiring maximum bearing support.

The following performance characteristics of epoxy and cement grouts listed below are for the purpose of determining the use of the most suitable product for an intended use. One property or requirement of a grout is seldom the only determining use, two or more properties are generally considered before selecting a product. The grout requirements for each use should be determined by the importance of the installation and overall suitability:

1. Compressive strength
   High early strength: For ambient temperature conditions of 70°F epoxy grout can be expected to be superior to cement products. Most epoxy grouts will produce 8,000–10,000 psi within 24 h as compared to most cement grouts producing only 3,500–4,000 psi @ 24 h.
   Ultimate: Epoxy grout produces 12,000–15,000 psi @ 7 days (standard maximum curing time) as opposed to cement grouts producing 5,000–7,000 psi @ 7 days and 8,000–10,000 after 28 days of proper curing.
   Conclusion: Both types of product produce higher ultimate compressive strength than required for typical construction grouting purposes. For high early strengths, epoxies have a commanding edge in strength gain at 70°F.
2. Lowest "downtime" (maintenance)
   As epoxy grouts produce very high early strengths their use may result in less production "downtime" for machinery in regrouting maintenance.
   Hydraulic cement grouts may be utilized providing less downtime if warm product and/or external heat is applied, and the same may be said for epoxy products under cool conditions. Unfortunately, cement grouts will not develop a good bond to steel.

CONCLUSION: On the basis of strength alone, epoxy grout has a distinct advantage over Portland cement grouts.

3. Chemical resistance
   There is no cement grout which will compete with epoxy for chemical resistance. Epoxy grouts are especially effective in applications that require resistance to dilute

acids. Exposed cement grouts may be coated for protection with epoxies or epoxy mortars, their effectiveness dependent on whether cracks develop in the coating exposing the substrate.

4. Volume change

Epoxy grouts indicate small amounts of shrinkage during cure. Expect approximately 90–95% plate bearing contact with epoxy grout. This figure is attributed essentially to entrapped air which rises to the interface of the grout and underside of the baseplate. This is accepted as normal and acceptable within industry.

Epoxy grout does not suffer drying shrinkage as such, but there is some chemical shrinkage when curing takes place. This is usually 0.5–1% of the total volume of the grout pour. This change in volume occurs in the pour or unrestrained areas.

Because of its high vibration damping characteristics and strength experience has shown that for many epoxy grouts as little as 75% contact has provided long-term service for most applications.

Portland cement grouts. Most cement grouts are designed to perform in accordance with The Corps of Engineers Specification For Nonshrink Grouts (CRD-C 621), and as a consequence must increase in volume in the *hardened* state when unconfined or go into compression when confined. In some critical alignment installations, this compression could result in loss of alignment by baseplate deformation.

CONCLUSION: Nonshrink cement grouts should conform to specification CRD-C 621. Epoxy grout should be used for applications requiring corrosion resistance, high early strength and a bond to concrete stronger than the concrete tensile strength.

---

5. Creep

Epoxy grouts are formulated to be highly creep resistant when tested in accordance with ASTM C-1181. Epoxy grouts are not as rigid as Portland cement grouting products. Load, temperature, and application must enter into this portion of the decision-making process. Do not base your decision solely by looking at published figures on creep testing.

6. Thermal coefficient of expansion (COTE)

The thermal coefficient of expansion of epoxy grouts will vary with composition of the product. This figure should closely match that of concrete and steel. The lowest reported COTE reported and verified so far is 11.2×10^{-6} in./in./°F.

The thermal coefficient of expansion for Portland cement grouts will be approximately 5.5×10^{-6} in./in./°F.

Where epoxies are used with concrete (or steel) the difference in COTE should be recognized and accommodated for with expansion joints and other techniques outlined in this book. This difference in COTE poses no grout problem.

7. Toughness

Epoxy grouts exhibit greater durability, strength, and overall toughness than cement grouts.

8. In-place costs—jobsite

The costs of materials, foundation, plate and forming preparations, mixing and placing are considerably higher for epoxy grouting than for cement grouts. Cleanliness and foundation moisture is more critical to the proper application of epoxy grout than for Portland cement grouting products. The long-term equipment reliability resulting from use of epoxy grouts are worth the extra effort and cost.

9. Damping

The term "damping" refers to the energy absorbing capacity in cyclic loading such as vibration and pounding. Epoxy grouts offer excellent "damping" characteristics superior to that of cement grout products.

10. Modulus of elasticity

The modulus of elasticity for epoxy grouts is in the magnitude of 1.5–2.5 million pounds. Epoxies can be formulated to affect a range from 0.5 to 4.5 million.

Cement grout can have a modulus of elasticity of 4.5 million at the 10,000 psi compressive strength level. This is dependent on how much water is applied when it is mixed and how it is cured.

11. Ranges of temperature limitations

Epoxy grout should not be placed when grout and foundation temperature cannot be maintained above 55°F or more and care must be exercised with placement at elevated temperatures. Epoxy grouts are not normally recommended for applications where ambient or operating temperatures exceed 180°F. Special formulations will withstand temperatures up to 250°F but tend to be extremely brittle and prone to crack.

Portland cement-type grouts may be placed as low as 40°F and with a recommended upper placing limit of 100°F, using due care particularly at lower limits. They may be utilized in working environments after thorough curing and then drying up to 400°F.

12. Thickness limitations

Some epoxy grouts can be poured in extremely large areas up to 7 ft × 7 ft × 18 in. in depth. This results in epoxy being used as a foundation reconstruction material to reduce downtime on regrouts.

Cement grouts may be placed in lifts at any thickness up to 12 in., or more, when foundation, bolts, rebar, and plates act as heat sinks.

13. Shipping and storing

Most epoxy products are not affected by freezing during shipping or storage. Prior to application at a jobsite the material should be preconditioned to ensure proper blending and thermal reaction of the resin and hardener. Environmental control should be used to keep the epoxy grout within a moderate temperature range during its placement and cure.

Cement grouts like epoxy are not affected by freezing temperatures. A recommended mixing and placing temperatures of 40–100°F without preconditioning or environmental control can be observed.

14. Curing

Epoxy grouts require no curing media other than a temperature above 55°F.

Cement grouts require early application of clean water saturated rags on exposed surfaces for 5 or more hours, followed by a curing compound application.

CONCLUSION: The performance characteristics of epoxy and cement grouts listed above are for the purpose of determining the use of the most suitable product. One property or requirement of a grout is seldom the only determining factor; you should consider two or more properties before selecting a product. The grout requirements for each use should be determined by the importance of the installation and overall suitability.

Merits of Epoxy Grouts Versus Cement Grouts

- Epoxy grout develops a greater bond to steel than does cement grout.
- Epoxy grout develops a higher bond to concrete than the concrete tensile strength (concrete tensile strength is approximated at 10% of compressive strength).
- Epoxy grout will develop a compressive strength greater than the concrete compressive strength between 24 and 48 h after placement.
- When properly mixed and applied, epoxy grout will provide a bearing area greater than 90%.
- Epoxy grout is 100% prepackaged.
- Epoxy grout has a high chemical and oil resistance, much greater than cement grout has.
- Epoxy grout will accept higher dynamic and static loading than will cement grout.
- Epoxy grout will make the equipment and the foundation a monolithic unit.

Cementious Grouts: Cement Grouting in Real Life

- The temperatures at which cement grout can be applied can range from less than 32°F to more than 110°F, with no special precautions necessary to ensure proper application.
- There is usually no qualified inspector present or any grout testing performed at the jobsite during the actual grouting.
- A general laborer or cement finisher will place the grout, so assume the following:
  - He will not read instructions on the bag.
  - He will add whatever water is required to get grout in place rather than tell his boss he can't place the grout because it is too stiff.
  - If the weather is hot, the laborer or cement finisher will retemper the grout with water to extend its working time.
  - He will not throw away grout that is unacceptable for use.
  - He will not tell the boss it's too cold to grout.

With this in mind, let's consider the following to be a general guide to cementious grouting.

Concrete Surface Preparation

Cleaning, roughening, and presoaking the concrete substrate with water for 24 h prior to actual grouting is essential before placing the cementious grout.

Roughening the concrete surface prior to grouting will ensure a proper bond of the grout to the substrate. The surface should be roughened to remove all laitenance as shown in Figure 3.1 and to expose sound concrete, particularly when dynamic, shear, or tensile forces are anticipated. The concrete surfaces should be chipped with a *chisel point* in a handheld chipping gun (Figure 3.2) to a roughness that produces aggregate exposure. Fifty percent of that exposed aggregate should shear, yet remain embedded in the concrete.

Presoaking will prevent the dry concrete from absorbing water out of the grout mixture before it has a chance to set. A dry concrete substrate could cause shrinkage

Figure 3.1 Properly chipped concrete.

Figure 3.2 Chipping gun.

even when a good grade of nonshrink grout is used, particularly when the grout is being placed in a stiff or dry-pack consistency. Don't take a chance: Always presoak the concrete and remove the excess or standing water just prior to placing grout.

Always shade the foundation from direct summer sunlight for 24 h before grouting and for at least 24 h after grouting. This will prevent the foundation from picking up too much heat and accelerating the grout's exothermic reaction. Also, keep the foundation covered after grouting to allow the grout to hydrate more slowly.

Steel surfaces should be free of dirt, oil, grease, mill scale, rust, or other contaminants.

Forming

When forms are required they should be sturdy, liquid-tight, nonabsorbent, and braced sufficiently to prevent any deformation. They should allow a 2 in. minimum space around the edge of the baseplate. Depending on the configuration of the baseplate, one side of the form is usually designated as the placement side. In some cases, the forms on the placement side should be spaced further out from the item being grouted to allow for a head box or flow box in which the grout will be placed and channeled under the item to be grouted. The top of the forms should extend a minimum of 1 in. above the bottom of the equipment being grouted.

Temperature

Follow the same guidelines for grouting as for concreting in cold or hot weather. Under cold weather conditions, proper preparation includes warming the concrete substrate and item to be grouted to a minimum of 45°F, storing the grout in a warm area, and using warm (up to 90°F) mixing water. In hot weather, the most common steps include using iced or cold mixing water and, if possible, storing the grout in a shaded or cool area. Cooling the baseplate with cold water is also advisable. In any case, always take steps to moderate temperature extremes.

To follow standard grouting procedures, keep the temperatures of the foundation, baseplates, mixing water, and grout at the following levels:

	Minimum (°F)	Preferred (°F)	Maximum (°F)
Foundation and baseplates	45	50–80	90
Mixing water	45	50–80	90
Grout at mixed and placed temperature	45	50–80	90

When grouting in cold weather or at the minimum temperatures, take care that the temperature of the foundation, the item being grouted, and the grout does not fall below 45°F until after the grout has taken its final set. The grout must be protected from freezing until it has reached at least 4,000 psi compressive strength.

Mixing Cement Grout

Generally, grout is mixed in a paddle-type *mortar mixer* (Figure 3.3) not in a rotary drum *cement mixer* as shown at the beginning of this chapter. This is because in a drum-type mixer with a rotating drum, centrifugal force pushes the grout to the outside of the drum, away from the center. Because the mixing blades are mounted to the drum and rotate with the drum, mixing is less than satisfactory. To improve the quality of the mix, the rpm of the cement mixer should be slowed down. Depending on the condition of the machine, the necessary rpm may be below its stall speed. Have more than one mixer on site to ensure continuous mixing and placement in the event of a mechanical breakdown. Mix grout until a uniform, lump-free consistency is obtained (usually 3–5 min is sufficient). Always follow the manufacturer's recommendations when mixing grout. And, if using a mixer powered by a small engine have extra fuel on site.

Figure 3.3 Mortar mixer.

Mix by placing the estimated amount of water into the mortar mixer and then slowly adding the dry grout. Always use clean drinking water. The amount of water needed depends on mixing efficiency, material, and ambient temperature conditions. Adjust the water to achieve the desired flow. Always follow the manufacturer's recommendations for the amount of water for the required consistency of the grout. Mix the grout per the manufacturer's instruction; do not overmix or undermix. Do not mix more grout than can be placed within the working time of the grout. Do not retemper the grout by adding water and remixing after the grout stiffens.

NOTE: Every effort should be made to minimize the transporting distance.

Placing and Curing

The cement grout is now mixed and ready to be placed. The key words now are *"Place continuously and quickly."* Begin the placement and continue it, if possible, from one side only. This will prevent entrapment of air or water beneath the equipment and avoid cold joints. The use of vibrators, straps, and rods in order to help move the grout may be a good idea when placing stiffer grout consistencies. However, in the case of highly flowable or fluid grouts, great caution should be exercised. External vibrations may cause bleeding and segregation of a fluid grout or may entrap air. If possible, use a grout with an extended set time to assist placement. Place fluid and flowable grout at least ½ in. above the bottom of the item being grouted to ensure complete filling of the grout space. After placement and the initial set, trim the surfaces with a trowel. Be sure the grout offers stiff resistance to penetration with a pointed mason's trowel before removing the grout forms or cutting back excessive grout. Cover the exposed grout with clean wet rags and maintain this moisture for 5–6h. Cure all the exposed grout with an approved membrane curing compound immediately after the wet rags are removed to further minimize the potential moisture loss within the grout.

Cement Grouting in Hot Weather

When using standard grouting procedures, high ambient and equipment temperatures tend to accelerate stiffening and require grout to be placed in a shorter period of time than is normally required under ideal conditions.

The preferred practice for hot weather grouting is to use cold materials and cool foundation and baseplates to extend the length of time the grout can be worked. This approach does not affect the nonshrink and strength development characteristics of the grout.

For best results when grouting in hot weather, try the following recommended temperature guidelines:

	Maximum (°F)	Preferred (°F)
Foundation and baseplates	70	80
Mixing water	70	80
Grout at mixed and placed temperature	70	80

- *Material storage.* Store the bags of grout in as cool a place as is practical (at least store them in shade). Remove the plastic wrap from the pallet of grout to help cool it, unless it is likely to be exposed to wet conditions.
- *Wetting the foundation.* Give extra attention to saturating the concrete base for 24h prior to grouting. Heat and wind cause rapid evaporation; therefore, the concrete base should be wetted liberally and frequently to prevent drying. Consider the use of wind breaks and sunshades to help maintain saturation.
- *Cooling the baseplate.* This can be accomplished while saturating the concrete base by covering both with wet burlap or cloth and keeping it wet.

Keep the temperature of the grout as mixed under 70°F preferably between 50°F and 55°F. The *as mixed* temperature is the temperature of the grout immediately after mixing.

RULE OF THUMB: Try to have the as *mixed* temperature of the grout at least as much under 70°F as the baseplate and foundation are above 70°F.

Minimum as mixed temperatures: The minimum allowable temperature for grout immediately after mixing or "as mixed" is 45°F. These minimums may be used only when the temperature of the baseplate and concrete foundation are measured, not estimated, and are found to be sufficiently warm.

- *Cooling the mixing water.* To lower the *as mixed* temperature of the grout, use cold water. If necessary, float ice in drums of water, using enough drums so when water is drawn off for mixing, the replacement water has time to cool. The use of a siphon or valve at the bottom of the drum facilitates drawing off the coldest water. Makeup water is added to the top of the drum where the ice floats. Insulating the drums or wrapping them with wet rags will help keep the water cold.

If large batches of grout are to be mixed, or if the packages of grout product are warmer than 90°F, consider substituting shaved ice for some or all of the mixing water on a pound-for-pound basis. Generally, shaved ice can be used in place of 50–75% of the mixing water by weight. Do not use more ice than will be completely melted within the proper mixing time of the grout. Unmelted ice poured with the grout will float to the top of the grout and melt, producing water pockets under the baseplate, resulting in a loss of bearing. Always pour the mixed grout through a ½ in. (or smaller) screen to remove unmelted ice, lumps, and foreign material.

It is good practice to take the temperature of the initial batch to determine whether more or less cooling is required.

- *Lowering mixer temperature.* If the mixer is warm, cool it by charging the mixer with cold or iced water to help reduce heating of the grout.
- *If the grout is being pumped.* A warm pump line can heat the grout and cause plugging. Covering the line with cloth or burlap kept continually wet will help cool the pump line. Also, consider using reflective insulation around the line and erecting sunshades to shield the line from the hot sun. The pump line can be cooled by filling it with chilled water or chilled cement slurry before batching the grout. However, the chilled priming mix must be completely discharged and discarded before pumping the grout.

When Cooling Cannot Be Accomplished

You should consider two approaches to cope with rapid setting in hot weather.

1. Form the area to be grouted into small sections so each section can be grouted individually. This is not a widely accepted approach, but it may lend itself to special types of grouting jobs.
2. Provide increased mixing capacity so that the grout can be poured faster and continuously. The less you can do to control temperature, the more rapidly you should mix and pour the grout under hot weather conditions. Consider using additional mixing equipment and locating the mixers so that the grout can be discharged through an inch screen directly onto the 45°-sloped form without transporting. Always provide straps positioned below the plate prior to grouting to aid in working as needed.

Cement Grouting in Cold Weather

The term *grouting* includes the preparation and assembly of carefully aligned plates that, in some cases, are shimmed and bolted into place. This can include machine baseplates, column plates, and structural elements that have been aligned and supported by a total-bearing, cementious grout. The successful installation of the cement grout depends a great deal on the environment in which it is poured. The following cold weather procedures are presented to aid in the successful installation of cementious grout under less-than-ideal conditions.

Recommended Temperature Guidelines for Cold Weather

	Maximum (°F)	Preferred (°F)
Foundation and baseplates	50	70
Mixing water	50	70
Grout at mixed and placed temperature	50	70

Mixed Grout Temperature

The temperature of the mixed and placed grout is affected by the temperature of the unmixed grout in the package, the temperature of the mixing water, the size of the batch being mixed, the temperature in the mixing and working area, (ambient temperature), and the temperature of plates and substrate.

- The optimum storage temperature for dry grout in cold weather is more than 60°F.
- Mixers and pumps (if used) should be warmed by rinsing them with hot water just before use. However, discard all warm rinse water before mixing.
- Do not use mixing water more than 90°F.
- Early-age strengths at cool temperatures are low, but grouts placed and cured at cool temperatures will be approximately as strong as grouts placed at normal temperatures after 28 days.
- Never mix grout to a consistency that produces bleeding or segregation under job conditions.

Foundation and Equipment Temperatures

- Accurately measure the temperature of the baseplate and the concrete foundation by using a surface thermometer. If you use an air or immersion thermometer, cover it with a piece of dry insulation material or dry rags.
- If the temperature of the baseplate or foundation is less than the 45°F minimum, warm the foundation to bring it up to the minimum. Heating large spaces to warm bedplates and foundations is slow and difficult. Use of infrared heaters is most effective because this heat penetrates solid objects.

Another method of heating baseplates is to either drape equipment with plastic sheeting (Figure 3.4) or construct a plastic-covered frame around the work area,

Figure 3.4 Environmental control is essential to successful grouting.

using space heaters inside. Small enclosures may be heated with forced hot air. Do not use open fires or kerosene heaters unless they are fully vented to the outside. In very cold weather, you may need to start heating 48–72 h in advance of grouting to bring the bedplates and concrete up to the minimum or desired temperature.

Curing Temperature

- Protect newly placed grout from freezing. After placement, maintain the grout at or above the minimum placing temperature until the grout has attained final set. Thereafter, keep the temperature above freezing until the compressive strength exceeds 4,000 psi.
- Early strengths may be accelerated by warm, moist curing. However, this heat must be carefully and uniformly applied to avoid thermal shock damage.
- Curing procedures to retain water for long-term strength gain and other properties are important, even in cool, moist conditions. Follow the recommendations on the bag.

Moisture Retention

- Moisture retention within the freshly placed grout should be extended when the in-place grout is below 50°F. This is important because the rate of hydration is slower at low temperatures.
- In order to assure optimum curing below 50°F, leave clean, moist rags on the exposed positions of the fresh grout for 24 h or more. Be sure to keep the rags moist during this period, covering them with plastic if necessary.
- Immediately after the removal of the moist rags and any trimming of the grout, paint the exposed areas with a double coat of a curing compound for long-term moisture retention and protection.

CONCLUSION: Read and follow the manufacturer's instructions and recommendations. Plan and discuss all critical grout work before starting the work. Everyone needs to know what is expected of them. Hot or cold, environmental control is necessary if the grout job is to be successful.

4 Epoxy Grout

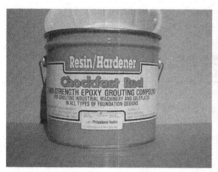

You get what you pay for, and sometimes less than you bargained for.

An Introduction to Epoxies

The first patent for an epoxy appeared in about 1936, but it wasn't until the early 1950s that epoxies became commercially available. It was about this same time that epoxies were first tried and evaluated in the construction industry. The initial results were encouraging and prompted additional testing and further work by the Department of Transportation as well as private companies. These early epoxy formulations appeared to have a number of excellent properties and demonstrated tremendous potential for solving some of the difficult problems of the concrete construction industry. Epoxies offered good adhesion to concrete, high bond strength, long-term resistance to harsh environments, easy application, and minimal shrinkage during curing.

Today, epoxies are formulated from a wide variety of resins and hardeners that result in extremely specialized products for different purposes and applications. Epoxies have achieved a respectable status and are the accepted standard in machinery installation.

Unfortunately, the construction industry today still base their product selection solely on cost. This inevitably results in the selection of an inferior product that is installed under high dynamic and static conditions, resulting in premature machine failure.

The Chemistry of Epoxy Grout

Epoxy grouts are usually composed of three components: the resin, normally called Part A; the curing agent, or hardener, often called Part B; and the aggregate system.

The Grouting Handbook. DOI: http://dx.doi.org/10.1016/B978-0-12-416585-4.00004-3

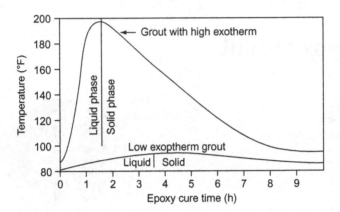

Figure 4.1 Epoxy grout exothermic curve.

Part A, the epoxy resin itself, is a product of a chemical reaction. The most common epoxy resins in use today are manufactured by reacting epichlorohydrin and bisphenol A. The properties of the epoxy resin can be altered either by changing the ratio of the reactants or by changing the processing procedures.

Part B of the system is the curing agent or hardener. Although there are a wide variety of agents that may be employed for curing epoxy resins, the most common are the amines, particularly aliphatic and aromatic amines.

When you mix the resin and hardener together, a chemical reaction begins that starts the hardening or curing process of the epoxy. The particular combination of resin, hardener, and other products determines the characteristics of the cured epoxy.

The epoxy molecule itself reacts again and again, growing in size, in a process called *polymerization*. This process of polymerizing and interlinking with surrounding molecules that are also polymerizing leads to the strong epoxy structure. The ratio of resin to curing agent may vary depending upon the particular resin and curing agent selected. Therefore, the resin and the curing agent combine in a very specific ratio; the proper quantity of each must be homogeneously mixed to ensure a strong bond. Unreacted resin or unreacted curing agent not only wastes material but also leaves weak links in the formation of the epoxy structure. In order to initiate the chemical reaction, each molecule of resin must come into physical contact with each molecule of hardener. If they don't, no reaction will take place. This chemical reaction is exothermic (Figure 4.1), and the amount of heat given off during this reaction varies significantly with different formulations.

NOTE: Following the manufacturer's directions for mixing. Suggestions on environmental control and placement procedures are critical.

General Characteristics of Epoxy Grouts

Temperature

Some of the most important factors influencing the rate of hardening, other than formulation, are the temperature of the substrate, the air temperature, and the temperature obtained by the mixed epoxy. As discussed earlier, the rate of hardening can be adjusted by preheating or cooling any of these parameters. In addition to influencing the rate of hardening, as you lower the temperature, you also observe an increase in viscosity of the material. If the ambient temperatures are 60–65°F, warming the material to 75–85°F gives much better flow but will normally give shorter pot life.

Chemical Resistance

Chemical attack is one of the more common causes of deterioration of concrete in industry today. Animal fats, natural and artificial oils, acids, alkalis, and various industrial salts are all damaging to concrete. Because of their high chemical resistance, you can formulate epoxies to protect concrete in these hostile environments. The degree and type of chemical resistance is related to the specific epoxy formulation and the choice of curing agent. Of the amine curing agents, aromatic amines are generally more resistant to chemical attack than aliphatic amines. However, aromatics offer relatively poor color stability and are usually quite viscous, which sometimes results in placement difficulties. As with many things in life, there is usually a trade-off when making decisions and designing grouting systems is no exception.

In reviewing potential formulations, the laboratory subjects the epoxies to discrete individual chemicals and environments. To check all formulations under all combinations of chemicals and environments would obviously be impractical. As a consequence, the laboratory can offer only very general guidelines on material suitability. Industrial environments may contain complex mixtures of acids, corrosive salts, and elevated temperatures. To determine the suitability of an epoxy in an acid environment, *test patches should be applied* and monitored (if practical) before the job is accomplished or performance levels are projected.

Note that if you are required to place epoxies over concrete previously exposed to chemical attack (particularly acids), be sure that *all* concrete exposed to that chemical is removed and that the remaining concrete is properly prepared before applying the material. Also, do not underestimate deterioration of concrete by *chemical vapors*. The attack may be quite deep due to concrete's permeability.

Surface Preparation

Many of the same commonsense precautions that apply to concrete also apply to epoxy grout. The concrete base should be clean and sound. You should remove all unsound concrete, laitenance, curing membranes, dust, dirt, and any other debris. Note that acid etching is not nearly as effective as mechanically abrading the surface for proper bonding. For metals, the surface must normally be sandblasted to white metal.

If oil or grease is present on the reinforcing bars, scrub the bars with a solvent that will not leave a residue. More details on surface preparation may be found in Chapter 5.

Mixing the Components

The Liquids

In mixing the liquid components of epoxy grout, remember that it is important to achieve a uniform and homogenous batch. The most common device for mixing the resin and hardener is a Jiffy Mixer Blade® (Figures 4.2 and 4.3). Using sticks, pieces of bent rebar, wire or paint mixing paddles cannot ensure a homogeneous mix of the material.

Figure 4.2 The Jiffy Mixer blade® can be used in either forward or reverse mode to eliminate sucking air into the resin.

Figure 4.3 Avoid vortexing of the liquids to minimize air entrainment.

The Aggregate

For large mixing jobs, combine the liquids and the aggregates in a mortar mixer. In some cases, a small amount of grout can be mixed in wheelbarrows. Whichever method is used do not over mix. Mix only long enough to completely wet out the aggregate.

For clean-up of equipment and personnel, follow the manufacturer's recommendations. Note that you should finish the clean-up before the epoxy is hardened. After the epoxy is hardened, it is extremely difficult to remove. Typical methods for removal of hardened epoxy are hammer and chisel or a grinder. You can use an acetylene torch to clean out hardened epoxy from mixing devices.

The Grout Didn't Get Hard!

Over my working career there have only been a few times when I was called in because the epoxy grout failed to harden.

I once got a call from a contractor who had poured about 50 units of epoxy grout and had two soft spots in it where the grout hadn't hardened. He made the pour about a week earlier and the daily temperature was around 90°F. Since I was close, I went to the job site to investigate and sure enough there were two areas about 6 ft apart that were not even close to being hard.

I asked the contractor if they had any material left over that they didn't use that I could get batch numbers from. We went to the onsite warehouse and there were 12 units of epoxy grout they had not used. A quick count showed there were two extra bottles of hardener that should not have been there. I asked the contractor how many people they had to mix and pour the grout and his reply was four.

Mixing and pouring 50 units with only four people meant someone was stretched thin. That someone turned out to be the guy mixing the grout. He was required to open the bags of aggregate, mix the resin and hardener, add the aggregate to the mortar mixer, pour the mixed liquid mixture into the mixer, mix, and pour it into the waiting wheelbarrows. All this and keep up with two wheelbarrows. Somewhere in the process of all this he lost track of adding the hardener to the resin and mixing. The end result was soft spots in the grout.

While not common this does happen from time to time. Most epoxy grout is packaged with the liquid resin and the liquid hardener in different containers as shown in Figure 4.4. Figures 4.5 and 4.6 show a different concept in liquid resin packaging.

Terminology

Heat Deflection Temperature

To understand the structural behavior of an epoxy in service, you must understand the concept of the glass transition temperature (the temperature at which the polymer

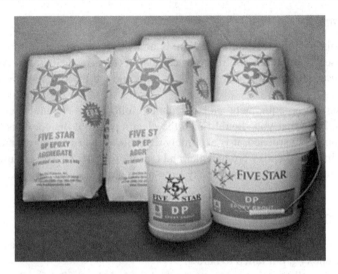

Figure 4.4 A typical three component epoxy grout unit.

Figure 4.5 A three component unit of epoxy grout packaged with a hardener tray in the resin container.

changes from a glassy to a flexible state). A value closely related to the glass transition temperature and used by construction designers is the *heat deflection temperature* (HDT), as measured by ASTM D-648. The HDT is the temperature at which an epoxy beam deflects 0.01 in. under a constant load of 264 psi. The beam (½ in. × ½ in. × 5 in.) is placed in an oil bath. The bath temperature is elevated slowly until the beam deflects. At the point where it deflects 0.01 in., the temperature of the bath is the polymer's HDT in degrees Fahrenheit.

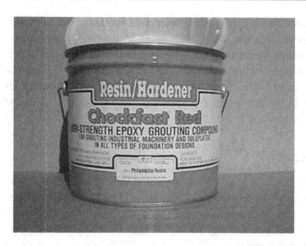

Figure 4.6 A plastic tray found in this epoxy grout resin container ensures the hardener will find its way into the resin and not be missed.

The temperature at this transition is profoundly significant because within the transition zone, designated as tough, all mechanical and physical properties including bonding strength, chemical resistance, and modulus of elasticity undergo a drastic change.

A Unit of Machinery Grout

A unit of machinery grout is the packaging from the grout manufacturer. It usually consists of several bags of aggregate and one or two cans of the liquid components Part A and Part B (the resin and the hardener). Yield of this packaging averages 1.5–2 ft^3 of mixed product.

Pot Life

The pot life of the epoxy begins when Part A and Part B are mixed and ends when the epoxy becomes too thick to use. This is normally before the material changes from a liquid to a solid as shown in Figure 4.1. This length of time is greatly influenced by the temperature of the mixed components of the epoxy system. Additionally, the ambient temperature, the concrete temperature, and the temperature of the component being grouted have a significant influence.

The volume of the pour is just as important when talking about the pot life of the mixed epoxy. One tends to associate pot life with working or placement time of the epoxy (the larger the volume of the poured epoxy, the shorter the pot life). This is due to the heat build-up from the exothermic reaction. This heat further accelerates the epoxy reaction. In the case of large pours, the epoxy remains in a mass, and the heat given off during the reaction becomes trapped in the mass. This trapped heat further accelerates the reaction of the mix, thereby creating more trapped heat. This heat reinforcement leads to a much shorter pot life than if the same amount of mixed epoxy were applied in a thin layer, such as a 2 in. deep pour, with a high surface area and more surrounding material available as a heat sink. You can also extend pot life by cooling (preconditioning) the resin and hardener prior to mixing, but take care to stay within the temperature design limits stated by the manufacturer. Excessive cooling will result in thickening of the resin and may result in a thicker-than-desirable consistency after mixing.

Exothermic Reaction

The exothermic reaction is the amount of heat developed as the result of combining the epoxy resin and the hardener. As shown in Figure 4.1, this reaction can in some cases be extremely high and can contribute greatly to cracking problems that can begin to appear as soon as the grout begins to cool. These cracking problems will be discussed later in this book.

Coefficient of Thermal Expansion

This is a measurement that indicates the material's tendency to expand and contract during temperature fluctuations. The higher the number, the greater the movement of the material for a given temperature change. A major critical difference between epoxies and concrete is evident in a comparison of coefficient of thermal expansion (COTE) (Figure 4.7). Epoxies have a much higher COTE than concrete and respond to temperature fluctuations to a much greater degree. Concrete has a tendency to restrain the movement of the epoxy and causes severe shear stresses at the interface during temperature changes. Unless you apply the epoxy with this in mind, concrete failure at the grout interface can result. This is of particular interest for large, thick pours. The aggregate filling of the epoxy reduces the COTE, but the resulting COTE may still be significantly higher than that of concrete or steel. Improvements have

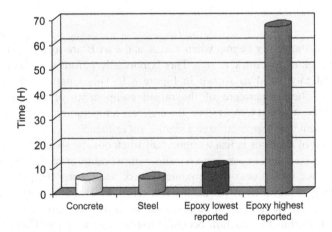

Figure 4.7 Comparison of COTE.

been made in the epoxies themselves that, with a little commonsense, can give quite satisfactory performance.

Creep: Is This a Problem in Epoxy Grout?

Creep is the tendency of a material, when placed under a continuous stress, to compress or move in a manner that relieves or reduces the stress. The rate of creep depends upon temperature and stress for a given epoxy formulation. The creep problem usually becomes more pronounced when the epoxy is subjected to elevated temperatures. Creep usually increases as temperatures rise. This usually does not present a problem in epoxy grout applications for several reasons.

ASTM C 1181 is the test for creep in epoxy machinery grouts and is probably the most confusing of all the ASTM tests for machinery grouts. Creep makes some epoxy formulations unsuitable for grouting machinery and for anchor bolt installations. The creep problem becomes more pronounced when the epoxy is subjected to elevated temperatures. *Creep increases as the temperature rises.* The *lower* the number, the better the creep resistance. Creep exists in every material known to human and affects all these materials in the same way. Creep in epoxy grout can be defined as follows: *Creep is the permanent deformation of the grout due to an applied continuous load that produces a level of stress in the grout less than the yield stress.*

Prior to the development of this test, most epoxy grout manufacturers used their own inhouse test for creep resistance. In 1991, ASTM Committee C-3 adopted a standard test for creep of epoxy machinery grouts. Some sketchy data exists on creep testing of other epoxy machinery grouts listed in accordance with ASTM C 1181. The following sections discuss this phenomenon and should give you a better understanding of creep and what this test really means.

All epoxy grouts don't creep the same way; the rate of creep for any epoxy grout depends on the following factors:

- Operating temperature of the grout
- Total load applied to the grout
- The formulation and manufacturing quality control (QC) of the product being tested.

QC is what makes some epoxy grout formulations unsuitable for grouting machinery and for anchor bolt installations. Therefore, in some cases, the creep phenomenon becomes more pronounced when the epoxy is subjected to elevated temperatures or loads. This is especially noticeable in high-flow grouts with low aggregate fill ratios.

In the mid 1970s, concern about creep started to emerge. It isn't clear who first decided that creep was the insidious problem that it is made out to be today, but, in any case, most epoxy grout manufacturers developed their own inhouse testing procedures to combat creep. The resultant data proved to be more confusing than the problem because everyone used a different procedure and obtained different results.

Beyond this, ASTM C 1181 does not call for any specific test temperature or load. It is left up to the person, or facility, performing the test to pick a number. Herein lays one of the problems that confuses most users of epoxy grout. This makes it tough to compare the various test results, sort of like comparing apples to lemons. Creep test result numbers are published for loads ranging from 250 to 800 psi.

Somewhere in the midst of time, the maritime industry decided that 500 psi was an acceptable load for epoxy chocks used under main propulsion engines. This 500 psi epoxy-chocking technology was brought to land-based engines and has been used successfully in the natural gas transmission industry up and down the pipeline for more than 20 years.

Some people would still like to believe that excessive creep is a function of the grout thickness. Actually, creep occurs only in the top 1–2 in. of an epoxy grout cap. This is because of the excellent thermal insulating capability of some epoxy grouts. The temperature at the surface of the grout dramatically decreases after about 1 in. of grout thickness. To date, no one has done any extensive field testing on deep pour epoxy grouts because of the man hours, equipment, and costs involved.

In 1983, a major chemical manufacturer and a major Texas university did some testing to "Predict Creep Lifetimes for Epoxy Grouts Under Integral Gas Compressors." They used five commercially available epoxy grouts and tested them for creep under laboratory conditions. This effort produced a six-page report that drew the following conclusions:

ASTM C 1181 calls for:
The test to be run over a 28-day period.
The test specimens to be donut shaped (2 in. thick × 4 in. in diameter with a 1 in. hole in the center).
4 Belleville Spring washers to be used to apply load to each specimen being tested. They should be capable of developing a load of approximately 5,800 lb when flattened.
Significant variations in creep behavior *were* observed among the various epoxy grouts. Varying anchor bolt tension, grout thickness, and bearing area of the chock or rail reduced *predicted* differential creep deformations.
Epoxy chock replacement, when the machine is epoxy chock mounted, is a relatively inexpensive way to correct alignment.

In 1997, the Gas Machinery Research Council (GMRC) performed research into creep for reciprocating compressor installations. The available GMRC reports are TR 97-5 "Epoxy Chock Material Creep Tests" and TR 97-6 "Compressor Anchor Bolt Design."

The name of TR 97-5 is a little misleading because it included not only epoxy chock testing but also epoxy grout. Testing was done on six different commercially available epoxy chocking and grouting products at elevated temperatures. This testing was performed under laboratory conditions (as is all epoxy grout physical property testing). The test cylinders for the grout phase of the testing were 4 in. × 8 in. and had thermocouples embedded in them to monitor temperatures. Their conclusion from the test results was that while creep occurs, the effect is rather *small* and *definitely manageable*.

Calculations run on epoxy grout creep definitely show that *in theory* creep can present a problem. However, these calculations assume many things, one of which is that the grout is heated *equally* through and through. In the real world, the surface temperature of the epoxy grout is the warmest—the material gets progressively cooler the deeper you measure. An old adage says that a bumble bee shouldn't be able to fly because it's not aerodynamically designed and the numbers prove it.

Unfortunately, no one bothered to tell the bee this "fact," so it goes on doing what science says it can't.

Deep pours of epoxy grout are the same theoretical calculations say that a certain effect can happen. However, just because brand X epoxy grout has a problem and can't be poured more than 6 in. deep without causing a problem doesn't mean brand Y can't be poured 18 in. deep without causing a problem. There is no easy way to actually test field installations because of the costs involved and their size.

At a technical presentation, I once heard a contractor who was a supposed expert in machinery grouting and repairing of concrete foundations say that he had observed some kind of support problem with *every* epoxy grout he had ever used. That seems to be a rather damaging statement because the only constant in that equation was *him*.

Have there been problems associated with deep epoxy grout pours? Sure, except the problems weren't with creep. The problems came from improper grout mixing and installation, poor equipment maintenance, weak concrete, loose anchor bolts, foundation settling, pipe strain, and other problems too numerous to mention.

It is difficult for someone to admit a mistake that required the machine to be regrouted or caused the foundation to fail. However, it's easy to say that mechanical and alignment problems are due to grout creep. The bottom line is this: No one has ever proven a machinery installation's grout failure was due directly to creep. Creep has been used as a catch-all phrase.

Deep, reconstructing pours of epoxy grout were first used by the owners of large pipeline compressors during regrouts. It was usually necessary to remove 18–24 in. of oil-soaked or damaged concrete to perform these repairs. This removed material was replaced with concrete that required from 14 to 28 days of cure. After the concrete was cured, the surface was chipped to remove laitenance and to provide a good surface profile for the grout to bond. This method often put the machine out of service from 6 to 14 weeks and usually resulted in a cold joint at the new concrete to old concrete interface.

To eliminate these long downtimes and designed-in failure points, the pipeline industry started using epoxy grout with pea gravel filler to form a rapid-set deep-pour compound. This significantly reduced machine downtime and eliminated the cold joint problem. As an extra benefit, these pioneers found that by using epoxy grout as a reconstruction material, they significantly reduced the machinery vibrations.

Conclusion

- ASTM C 1181, 2.1 states by its own admission that the results of this test cannot be prorated for different thicknesses and that no correlation has been established to *actual use conditions.*
- No qualified field testing confirms that the deeper the epoxy grout pour, the greater the creep phenomenon.
- No documented evidence proves that creep has ever resulted in the failure of an epoxy grout.
- Actual creep is minimal, is confined to the top 2 in. of the grout, and is less than a few thousands of an inch.
- In order for creep to become a problem, the grout must be subjected to a combination of high temperature and high compressive loads.

General Installation Procedures for Epoxy Grout

Generally, grout should be mixed as follows:

- Store all the grout materials in a controlled environment at 65–80°F for 48 h prior to mixing. If the grouting materials cannot be preconditioned, contact the grout manufacturer for their recommendations.
- If possible, the grout manufacturer or their approved representative should be allowed to inspect the foundation and witness the grout mixing and pouring procedure.
- The grout should be mixed in a clean, slow speed (15–20 rpm) portable mortar mixer in good condition.
- The epoxy grout liquid resin and hardener should be thoroughly mixed for 3 min with a medium Jiffy Mixer Blade® at a speed of no greater than 200–250 rpm.

CAUTION: Vortexing or whirlpooling of the liquid components during this phase of the mixing will create air bubbles and may cause surface imperfections to appear in the grout or voids under baseplates.

- Pour the resin and hardener mixture into a mortar mixer. Add one bag of aggregate at a time.

Mix only long enough to thoroughly wet the mixture.

NOTE: Cutting back a half a bag of aggregate on the first batch to wet out the mixer is a standard procedure. All other batches should use all aggregate bags of the factory proportions per unit of grout.

Also, mixing time of epoxy grout is dependent on ambient and material temperatures.

Concrete Temperature

The concrete temperature should be a minimum of 65°F and a maximum of 90°F unless approved by the grout manufacturer. If the concrete temperature is below 65°F or above 90°F, build a temporary environmental control structure around the machine.

Ambient Temperature and its Effect on Epoxy Grout

As discussed earlier in this chapter, the ultimate strength developed by an epoxy grout system occurs as a result of a chemical reaction between the resin and the hardener. As with most exothermic chemical reactions, this process occurs more rapidly at higher temperatures and more slowly at lower temperatures. This means that the strength developed by an epoxy grout proceeds slowly at 55°F. On the other hand, as the temperature of the system is increased by an external heat source, the rate of the chemical reaction increases. This causes the rate of the strength development to increase also.

Temperature also affects grout viscosity. As the temperature lowers, both the resin and the hardener thicken. At higher temperatures, they thin out. This means that at 90°F, the two epoxy components combine readily and are very easy to mix together. Around 55°F, the liquids are much thicker and much more difficult to mix. At cooler temperatures, you may need a longer mixing time than at warmer ones to be assured of achieving a uniformly mixed solution.

This change in viscosity with temperature also affects the ease of placement of the grout. At higher temperatures, the grout flows very easily. However, at 55°F, epoxy grouts are much more sluggish. More effort is necessary to move the grout through the head box, as the grout tends to cling to the head box walls and concrete very strongly.

A grout job started late in the evening during cool weather may end up taking a few hours longer than expected to complete. The cooler temperatures after the sun sets will usually make the grout placement proceed more slowly.

With this in mind, follow the instructions given in the following sections.

Epoxy Grouting in Cold Weather

Handling and Storage
- Store all components in a dry and weatherproof area prior to grouting. Under no circumstances should you store grouting components outside or in an area that cannot be heated to 65°F or above.
- For optimum handling characteristics, adjust all components (particularly the aggregate portion) to a temperature of 65–80°F prior to grouting. Do not stack aggregate bags—spreading them out allows for equal heating.

Preparation
- Precondition the work area, including foundation and machinery, to a temperature greater than 65°F for 24 h prior to grouting. This can best be accomplished by constructing a temporary structure around the work area with a suitable covering, if necessary.
- Ensure that the temperature of the concrete foundation and steel machinery is a minimum of 65°F prior to grouting.

Placement
- Coordinate grouting for minimum placement time.

Curing
- Hold the work area, including foundation and machinery, at a minimum temperature of 65°F for 48 h after placement of the grout.
- Do not position heating sources (high intensity lamps, steam, or forced air heaters, and so on) so as to create hot spots (localized heating) on the grout.
- After the grout is fully cured, equalize the temperature inside the temporary structure with the external temperature gradually.

Epoxy Grouting in Hot Weather

Handling and Storage
- Store all components in a dry and weatherproof area prior to grouting.
- Under no circumstances should you store them outside in direct sunlight or under a tarpaulin.
- For optimum handling characteristics, resin and hardener components shall be preconditioned to a temperature of 65–80°F prior to grouting.

Preparation
- Protect the work area, including foundation, machinery, and mixing equipment, from direct sunlight prior to grouting. This can best be accomplished by a temporary cover around the work area, if required.

- Test the temperature of the concrete foundation and machinery using a surface thermometer prior to grouting. Surface temperatures shall not exceed 90°F.

Placement
- If you expect ambient temperatures greater than 90°F, grout during early morning or evening hours when the temperature is lower.

Curing
- If you expect ambient temperatures greater than 90°F, protect the work area, including foundation and machinery, from direct sunlight after placement of the grout, until the grout has cured and returned to ambient temperature.

Epoxy Grout Flow Versus Clearance

You can't have a high-flow epoxy grout at a low price.

Over the years epoxy grouting technology has made tremendous advances. From its beginning in the early 1950s till today, we've seen the possible depth of an epoxy grout pour rise from 1 in. deep to greater than 2 ft deep. The grouting products of today are more crack resistant and generate a lower exotherm than their predecessors did. Unfortunately, with all these technological advances, most grouting specifications have not been modified or corrected to reflect these changes. The one item that has been most affected is grout flow versus clearance.

When using a conventional epoxy grout, most engineers call for a clearance of usually 1–1½ in. between the foundation and the equipment to be grouted. This is fine when using a high flow-type epoxy grout; however, when using a flowable epoxy grout, you may encounter problems when flowing distance is greater than 2 ft or when cooler temperatures are encountered.

As epoxy grout flows across concrete, it gives up a percentage of its resin to the concrete. To compensate for this, the grout installer usually reduces the prepackaged, premeasured aggregate to obtain a more fluid mix.

The following changes occur when this reduction of aggregate is allowed:

- The physical properties of the epoxy grout are reduced.
- The exothermic reaction of the grout is increased.
- The cost of the installation is increased.

Aggregate is the least costly of all the components of epoxy grout but probably the most important. Increasing the clearance space under a piece of equipment to be grouted is less expensive than reducing the aggregate filler to improve flow. With this in mind, the following is presented for your consideration.

Flow Under Large Sole Plates, Rails, or Skid-Mounted Units Using Flowable Epoxy Grout

2 in. of clearance required for the first 2 ft of distance.
1½ in. of additional clearance is required for each additional feet of distance up to 8 ft.

69–78°F (Material and Base)

Distance (ft)	Clearance (in.)
2	2
3	2½
4	3
5	3½
6	4
7	4½
8	5

For cooler temperatures (55–68°F), increase clearance by ¾ in.

Distance (ft)	2	3	4	5	6*	7*	8*
Clearance (in.)	2¾	3½	4¼	5			

*Skid-mounted equipment can be grouted with lower clearance (2–3 in.) by using special techniques.

For warmer temperatures (79–90°F), a good rule of thumb is to increase the grout space by ¼ in. of clearance for every foot above 2 ft.

Distance (ft)	2	3	4	5	6	7	8
Clearance (in.)	2	2¼	2½	2¾	3	3¼	3½

5 Selecting an Epoxy Grout

The bitterness of poor quality remains long after the sweetness of low cost has been forgotten.

The selection of an epoxy grout for use on a major project often seems to be left in the hands of the buyer. Most material specifications for grout written today say something to the effect of "or equal" after the listing of approved products. The engineer might as well write "use the cheapest product that can be found," into the specification. Not all machinery grout manufacturers market their products through distributors who provide the onsite assistance as well as the product. Some of these distributors sell the grout at list price or give a quantity discount. Their technical service is built into this price. When the price must be cut to the bone, the service is the first thing that gets cut. Later, when problems arise and someone is needed at the job site, it's a difficult process to have funds appropriated for the necessary service work.

Estimating the number of epoxy grout manufacturers in the world today is difficult. Their numbers fluctuate and include regional manufacturers and those who manufacture their products in their garages. Becoming an epoxy grout manufacturer is not difficult. All you need to do is contact a local chemical manufacturer and ask for an epoxy resin and hardener system. Add some aggregate, and you have an epoxy grout. Sound too easy? It is.

Two types of epoxy grout systems exist: general construction epoxy grouts and those designed specifically for machinery installations. General construction grouts can be classified or recognized by their depth of pour limitation (1–2 in.), high exothermic reaction, and high coefficient of thermal expansion (COTE).

The Grouting Handbook. DOI: http://dx.doi.org/10.1016/B978-0-12-416585-4.00005-5

The following physical properties are those of machinery grouts. They have been manufactured continuously for more than 25 years and are available worldwide. They are tested in accordance with tests developed by ASTM Committee C3, which has jurisdiction over epoxy machinery grouts. Their technology and the ability of these products to make deep reconstructing single pours of 18 in. or greater set these products apart. The following ASTM test methods and resultant physical properties of the major epoxy grouts used to install machinery were derived from published and independent test data of those manufacturers.

ASTM C 307: Standard test method for tensile strength of chemical-resistant mortar, grouts, and monolithic surfacing (uses standard "dogbone" tensile molds).

Chockfast Red SG	3,000 psi
Chockfast Red	2,460 psi
Ceilcote 648CP + Std.	2,200 psi
Escoweld 7505E/7530	2,100 psi
Ceilcote 648 CP + HF	2,000 psi
Five Star Epoxy Grout DP	1,700 psi
Five Star Epoxy Grout	1,500 psi

ASTM C 579 Method B: Standard test methods for compressive strength of chemical-resistant mortars, grouts, and monolithic surfacing using a 2 in. cube at room temperature.

Chockfast Red SG	18,121 psi
Five Star	16,000 psi
Chockfast Red	15,250 psi
Five Star DP	15,000 psi
Ceilcote 648 CP + Std.	14,000 psi
Escoweld 7505E/7530	14,000 psi
Ceilcote 648 CP + HF	11,500 psi

The compressive strength published is fairly impressive until one actually puts pencil to paper: You will see the real story then.

The published data from the technical literature published in March, 1996 of an epoxy grout manufacturer (not listed in our comparison) stated the following:

Test Temperature (°F)	Compressive Strength (psi)
73	19,700
120	11,200
140	7,500*
160	6,200
180	5,000
200	3,700

*Normal operating temperature of reciprocating compressors.

Once again, these figures are impressive until you run the numbers.

Test Temperature (°F)	Compressive Strength (psi)	Temperature Difference (°F)	Reduction in Strength
73	19,700	–	–
120	11,200	47	47.0% (8,500 psi)
140	*7,500	67	62.0% (12,200 psi)
160	6,200	87	68.6% (13,500 psi)
180	5,000	107	74.7% (14,700 psi)
200	3,700	127	81.2% (16,000 psi)

When you compare these numbers with one of the four products listed earlier, you see a distinct difference between a product with a higher compressive strength at ambient temperature and an operating temperature.

73°F	150°F	Temperature Difference (°F)	Reduction in Strength
15,250 psi	9,600 psi	77	38.5% (6,000 psi)
9,700 psi	7,500 psi	67	61.9% (12,200 psi)

This product has a greater reduction in strength for a *smaller* temperature differential.

CAUTION: Do not go by published compressive at laboratory test temperatures alone. Instead look at what the strength will be at machine operating temperature.

ASTM C 531: Standard test method for linear shrinkage and COTE, of chemical-resistant mortars, grouts, monolithic surfacing (uses 1 in.×1 in.×10 in. molds with studs).

	COTE (in./in.)	Shrinkage (in./in.)
Chockfast Red SG	10.8×10^6	0.00021
Chockfast Red	11.2×10^6	0.00002
Escoweld 7505E/7530	14.0×10^6	0.00036
*Five Star Epoxy Grout	15.0×10^6	0.0000
*Five Star Epoxy Grout DP	18.0×10^6	0.0000
Ceilcote 648CP + std	19.0×10^6	0.0005
Ceilcote 648CP + hf	23.0×10^6	0.00065

*Indicates that the test used was not designed for the material being tested.

* **ASTM C 827-87:** Standard method for testing change in height at early ages of cylindrical specimens from *cementious mixtures*. This test is mainly used for measuring gas expansion in the plastic state and is used by only one epoxy grout manufacturer. No data is published as to what pressure can be expected against the underside of a baseplate in a restrained or confined area due to the swelling that takes place. No other epoxy grout manufacturer uses this test *ASTM does not consider this test relevant to epoxy grout.*

ASTM C 580-85: Standard test method for flexural strength and modulus of elasticity of chemical-resistant mortars, grouts, and monolithic surfacing (uses a 1 in.×1 in.×10 in. bar).

	Flexural (psi)	Modulus (psi)
Five Star Epoxy Grout DP	5,200	2.48×10^6
Escoweld 7505E/7530	4,700	1.8×10^6
Ceilcote 648-CP + STD	4,500	2.1×10^6
Chockfast Red	4,180	2.07×10^6
Five Star Epoxy Grout	4,000	1.67×10^6
Ceilcote 648-CP + HF	4,000	1.6×10^6
Chockfast Red SG	3,737	1.7×10^6

ASTM C 1181-91: Standard test methods for compressive creep of chemical-resistant polymer machinery grouts.

Chockfast Red SG	2.0×10^3 in./in. at 140°F, 400 psi
Five Star Epoxy Grout DP	2.7×10^3 in./in. at 140°F, 400 psi
Chockfast Red	2.8×10^3 in./in. at 140°F, 400 psi
Ceilcote 648 CP + STD	4.0×10^3 in./in. at 140°F, 600 psi
Five Star Epoxy Grout	4.4×10^3 in./in. at 180°F, 400 psi
Escoweld 7505E/7530	4.6×10^3 in./in. at 140°F, 600 psi
Ceilcote 648 CP + HF	6.0×10^3 in./in. at 140°F, 600 psi

CONCLUSION: Basing a product selection on cost alone is a bad idea. Select a product that provides engineering assistance in the engineering phase of a project as well as onsite assistance. It's easier to solve problems in an office than it is in the field.

6 The Use of Rebar and Expansion Joints in Epoxy Grout

I am often asked about the use of steel-reinforcing rods, or rebar, in epoxy grout foundations. Because rebar has historically been used in concrete, it seems logical that it should also be beneficial in epoxy grouts, but this is not necessarily true.

As we discussed in Chapter 3, concrete, as a rule, has a tensile strength of only about 10% of its compressive strength. In other words, a 3,000 psi concrete in compression should have a tensile strength of approximately 300 psi. Steel rebar is used to add tensile strength to concrete members. Epoxy grouts, however, have considerably higher tensile strengths, usually in the range of 1,500–2,000 psi and should not require additional reinforcement in most applications.

As we discussed in Chapter 5, the principal concern with using rebar in epoxies lies in their different coefficients of linear thermal expansion. Or how much they will "grow" or "shrink" with changes in temperature. Concrete and steel have similar coefficients and are therefore compatible when used together. Epoxy grouts, however, have higher coefficients. Some formulations can have rates of expansion almost five times that of steel (Figure 6.1).

The Grouting Handbook. DOI: http://dx.doi.org/10.1016/B978-0-12-416585-4.00006-7

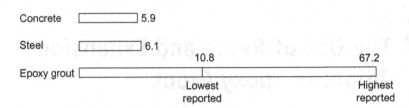

Figure 6.1 The coefficients of linear thermal expansion of concrete, steel, and epoxy grout.

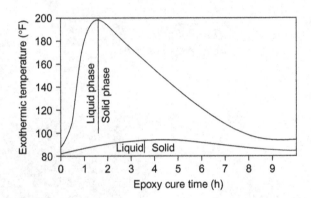

Figure 6.2 Peak exotherm of a typical epoxy grout versus a grout designed for deep pours.

Epoxies are exothermic, or heat-creating, in their curing process. Different epoxy grout formulations have different curing reactions that can vary from a peak exotherm of a few degrees to well over 100°F above their ambient pour temperatures. Because maximum exotherm is related to the size of the mass, the "hotter" epoxy grouts are limited to relatively shallow pours. You can obtain an indication of the amount of exotherm to be expected by referring to the manufacturer's maximum recommended pour depth; the greater the depth, the more gentle the cure.

Epoxy grouts go from a liquid to a solid state near their peak exotherm (Figure 6.2). When grout is poured on a concrete base with exposed rebar, the curing reaction heats both materials. As the grout solidifies, it can be at any temperature from warm to hot and encapsulates the steel rods. You can easily visualize what happens as the grout and steel cool to ambient temperatures and contract at different rates. Epoxy grout is put in tension when it contracts more than steel, which creates stress in the epoxy. The greater the temperature and thermal expansion differences are between epoxy and steel, the greater amount of stress in the grout (Figure 6.3). This can cause cracks in the grout that may appear shortly after it has cured. Later drop in temperature due to seasonal temperature changes can increase the stresses and cause cracks. Some grout manufacturers recommend the use of horizontal rebar in deep pours to act as a heat sink and reduce the peak exotherm of the epoxy grout. However, this may actually be the cause of cracks if the foundation ever sees

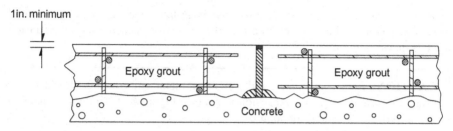

Figure 6.3 The placement of rebar and expansion joints in deep pour epoxy grout.

significant fluctuation in temperatures. This same phenomenon can also be the cause of loose soleplates or rails if they are set in an epoxy grout with a high coefficient of expansion.

Because the total thermal expansion and contraction of a material is directly proportional to its length, the mismatch between rebar and epoxy grouts applies primarily to the long horizontal rods commonly found in machinery foundations. Short vertical pins placed around the foundation perimeter provide a mechanical lock between the grout and the concrete. These pins will not usually precipitate a stress crack in a good quality epoxy grout, provided they are at least 1 in. in from any surface.

Having a coefficient of expansion as close as possible to concrete and steel is also very important in situations where considerable changes in temperature are possible. As the temperature rises, the grout tries to expand more than the steel if rebar is present. This puts the grout in compression and the rebar in tension, which is allowable because the strengths of these materials are very high under these conditions. The area for concern, however, is at the interface between the concrete and the epoxy grout. As the grout expands at a faster rate, it puts the concrete in tension, which, as discussed earlier, is not one of concrete's strengths. If the stresses are great enough, the grout will shear the concrete just below the bond line. You will have a harder time detecting this condition than a crack in the grout, but if it is subjected to dynamic forces, as under an engine, this horizontal crack will make itself evident in time.

This tendency for the grout to shear its bond with the concrete can, however, be minimized. Because the amount of thermal expansion is proportional to the length, properly spaced expansion joints in the grout reduce the effective length by segmenting and thereby minimizing stresses on the bond area. The greater the discrepancy between thermal coefficients of materials, the closer expansion joints must be to ensure a lasting structure.

Grout manufacturers publish various physical properties for their materials. Compressive strength, compressive modulus of elasticity, and tensile strength, although important, are overemphasized because they are far greater than concrete and usually loaded to a fraction of their limits. The most important design criteria, if an epoxy grout is to be used with other materials, such as concrete and steel, is compatible with these materials. Because very few environments are absolutely stable, you must calculate the effects of temperature changes and eliminate undue stresses by choosing proper materials.

If, for whatever reason, horizontal rebar is installed and epoxy grout poured around it, you should take care to prevent the rebar from penetrating an expansion joint as shown in Figure 6.3.

> **CAUTION:** Rebar, soleplates, rails, or other metal embedments should not bridge or pass through the expansion joints (see Figure 6.3). To do so defeats the purpose of the expansion joint.

Best Advice: Use vertical rebar only to strengthen the grout to concrete interface to reduce the possibility of edgelifting. Rebar should be at least 1 in. below the horizontal surface of the grout to reduce the possibility of a surface crack. Also, the rebar must be at least 3 in. from the vertical surface of the grout.

Edgelifting: Cause and Cure

The first indication of edgelifting is a crack just below the grout line.

Edgelifting caused by poor quality concrete and no rebar.

Preventing edgelifting is much easier than repairing it. Edgelifting, or curling as it is sometimes referred to, is the phenomenon caused by the difference in the rate of thermal contraction between epoxy grout and concrete with low tensile strength. Figure 6.4 illustrates how the different coefficients of thermal expansion/contraction react one to the other during a temperature decrease. The result of this differential contraction results in the tensile failure of the concrete just below the grout–concrete interface. As discussed earlier, the tensile strength of concrete is about 10% of its compressive strength. When you apply a sufficient amount of heat to the epoxy grout, the reverse occurs. The epoxy expands at a rate greater than the concrete and the crack closes. This is why these cracks are more noticeable after a sudden temperature drop, such as that first real cold front during the winter, than they are in the summer.

You can eliminate edgelifting in several ways, one of which is to install doweling around the outside edge of the foundation, as shown in Figure 6.5. This is normally done when the epoxy grout cap is poured 8–12 in. (or more) away from the machinery base and is done for cosmetic or sealing reasons rather than for equipment support.

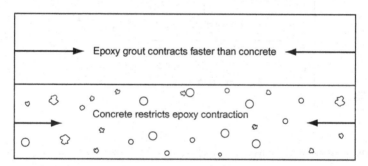

Figure 6.4 Different coefficients of thermal expansion/contraction react differently one to the other during a temperature change.

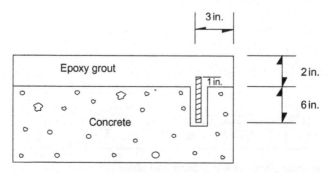

Figure 6.5 Use doweling around the outside of the grout cap to prevent edgelifting.

By changing the dimensions of the grout cap to a *depth greater than width* as shown in Figure 6.6, you reduce the affected area that would be subject to thermal contraction. However, by reducing dimension W to the same as or less than dimension D, you can reduce the possibility of edgelifting. Edgelifting will not occur where the epoxy grout is in compression. If you use this type of application, be sure that dimension W allows sufficient room for proper grout placement. A head box will be required to enhance flow under large machinery or plates when using a minimum grout shoulder. Reducing the amount of aggregate in the grout to enhance flow creates another set of problems.

Another way to prevent edgelifting is to radius or chip away the outside edge of the concrete foundation. There are several schools of thought on how much to chip away; however, all agree that a 45° angle is the best method and that 2–6 in. is a sufficient area. As shown in Figure 6.7, this radiusing of the concrete will usually

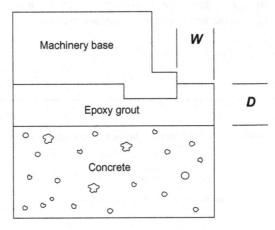

Figure 6.6 Keep the dimensions of the grout cap to a *depth greater than width.*

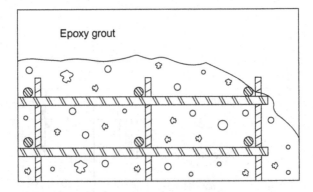

Figure 6.7 Expose peripheral rebar to prevent edgelifting.

expose rebar that was originally installed in the concrete. This exposed rebar will further aid in the prevention of edgelifting.

Why and When to Use Expansion Joints in Epoxy Grout

In order to maintain alignment of grouted equipment, most epoxy machinery grouts are designed to be rigid and have high resistance to creep. As a result, stresses developed during cure and subsequent temperature changes may result in cracking. Vertical cracks do not usually impair the grout's supporting capability; however, they are undesirable because of their appearance. Cracks in the epoxy grout can allow oil and water to migrate down to the concrete substrate and begin to deteriorate the concrete. Expansion joints should be used in all foundation designs more than 4 ft in length or width. Using expansion joints will also help to reduce the possibility of cracking on long grout pours where the difference between the maximum temperature the grout will experience or has experienced (due to peak exotherm or maximum operating temperature) and the minimum temperature the grout will experience will be below 50°F. In other words, always use expansion joints.

Suggested Expansion Joint Locations and Basic Construction Material

Expansion joints can be located every 3–5 ft, depending on the length and width of the foundation. They should be positioned so as not to interfere with soleplate, chock, or anchor bolt locations. For best results, *always* consult the grout manufacturer about their recommended expansion joint design and location.

The primary joint material should be ¾–1 in. thick. Soft wood or dense styrofoam make excellent expansion joint materials. They are resistant to water and oil and are easily compressible.

The Secondary Seal

After the concrete surface has been chipped and the forms erected, the expansion joints may be installed. A mixture of one part elastomeric expansion joint compound and approximately four to seven parts clean dry sand should be applied onto the concrete 3 in. wide and 1 in. thick along the area to receive the expansion joint. The expansion joint material is then pressed into this mixture to a depth of ½–¾ in. The expansion joint compound and sand is allowed to cure. This now becomes the secondary seal, as shown in Figure 6.8.

Figure 6.9 shows an alternate method to use when installing the secondary seal. Chip a groove 1 in. deep into the concrete. This groove should be about 1 in. wider than the expansion joint material. Depending on personal preference, the joint can be set before or after the forms are erected. Whichever method is used, the elastomeric

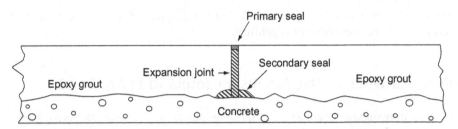

Figure 6.8 The secondary seal when cured provides support for the expansion joint when more grout is poured on one side than the other.

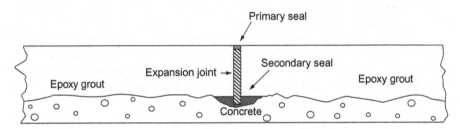

Figure 6.9 Alternate method for secondary seal.

epoxy should be mixed with a slight amount of sand (coarse or fine) so it remains fluid but is cost effective. After the elastomeric epoxy is in place, the expansion joint material is pressed into it, and it is allowed to cure.

The Primary Seal

On most expansion joints, the primary seal can be poured in place (Figure 6.10). This is accomplished by installing a strip of wood or other material wrapped in polyethylene or duct tape (to prevent the grout from bonding). It is installed so a portion of it extends above the grout when poured and down the vertical face. After the grout has hardened, these strips are removed from the horizontal surface and the vertical face. The void is then filled with elastomeric epoxy expansion joint compound without sand.

Mixing Elastomeric Epoxy Expansion Joint Compound and Sand

To form a nonslump workable paste, proper mixing is essential:

- Using a good quality mixer blade, mix the elastomeric epoxy resin and hardener for 3 min at approximately 250 rpm.
- Pour the mixture into a 5 gal bucket and slowly add clean dry sand while agitating with the mixer blade. Add only enough sand to form a nonslump, nonrun mixture (approximately a 7:1 ratio). This will vary depending on the grade of sand used. A KOL-type mixer can also be used to mix the elastomeric epoxy and sand.

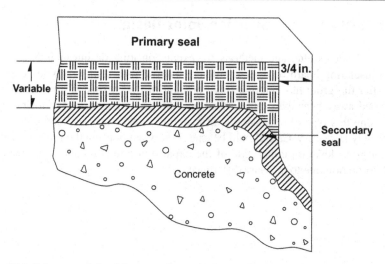

Figure 6.10 Primary seal should protect the vertical grout face as well as the horizontal surface.

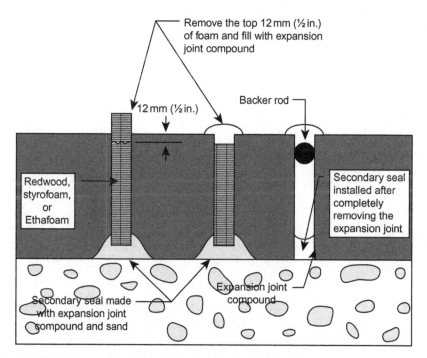

Figure 6.11 Various styles of expansion joints for machinery grouting.

Other Methods for Expansion Joint Design

Figure 6.11 shows alternate methods for expansion joint design include using styrofoam insulating board or rigid styrofoam material in lieu of a redwood expansion
joint. After the grout has cured, a small amount of the expansion joint material can
be chipped away near the grout surface and the gap filled with elastomeric epoxy.
When using this type of expansion joint material, a secondary seal along the bottom
is absolutely necessary. Care should be exercised when pouring the grout to maintain
an equal grout level on either side of the expansion joint to reduce the chances of
joint deformation due to pressure differences.

7 Cracking in Epoxy Grout

Only two things will cause cracking in epoxy grout: thermally induced stress and mechanically induced stress. A crack does not mean that the grout has failed. It simply means that the grout is relieving some type of mechanically or thermally induced stress. More simply put, the grout is showing where the expansion joints should have been.

Thermal Stress Cracks

Rapid temperature drops can have a serious effect on cured epoxy grout. Epoxy grouts can develop cracks in long or wide pours if they are subjected to a temperature differential greater than 50°F; that is, if, during a seasonal cycle, the grout will be subjected to temperatures more than 50°F cooler than when it was poured. In the equatorial regions, this does not present a problem because there is rarely a 30°F difference between summer and winter. However, in climates where a high temperature

The Grouting Handbook. DOI: http://dx.doi.org/10.1016/B978-0-12-416585-4.00007-9

differential of as much as 100°F or more can occur between summer and winter, thermal cracking can and will occur unless you take proper precautions.

Mechanical Stress Cracks

These cracks can also occur at a point where the stress concentration is the highest, such as unwrapped anchor bolts, square-cornered shim blocks, soleplates, inside right angles, or jackscrew pads. Mechanical stress cracks are also a result of thermal changes and appear where differential contraction between the epoxy grout and the steel embedments in the grout occurs.

Vertical Cracks

You should seal this type of crack (as shown in Figure 7.1) immediately to prevent oil or contaminants from penetrating the grout and reaching the concrete below. The sealant you choose depends on the type of service and the environmental conditions to which the grout will be subjected. If sealing the crack is mainly to prevent water accumulation and freeze–thaw damage, use a one component industrial-grade caulk such as room temperature vulcanizing (RTV) silicone. If there will be continuous presence of lubricating oil, spilled fuel, or other contaminants in the crack area, use a more resistant, elastomeric sealant. Most epoxy grout manufacturers have such a product that is used to seal expansion joints in epoxy grout pours.

To repair a vertical crack, try the following:

1. Chip or grind out the top of the crack by making ¼–½ in. V or leave it unchipped and build a ½ in. high dam of *duct seal* on either side of the crack. After creating the V, blow out the dust with oil and water free air and keep it dry.

Figure 7.1 Designed in stress concentration points.

> **NOTE:** When using the dam method, be sure to remove the duct seal and excess epoxy before it hardens. Duct tape applied along either side of the crack ahead of time will help in clean-up.

2. Use a two-part liquid epoxy injection grout or the two liquid parts from a conventional three-component unit of epoxy grout. This material will be thinner than an expansion joint compound or sealing caulk such as silicone. The liquid epoxy mixture can flow into the crack and will penetrate several inches.
3. If water is in the cracks, speed its removal and drying by pouring a fast evaporating solvent that doesn't leave a residue, such as acetone, methanol, or anhydrous alcohol into the crack. These will absorb the water and carry it out as they evaporate.

> **CAUTION:** Do not use a petroleum solvent that will leave a hydrocarbon residue. All epoxies and solvents are potentially hazardous. Follow safe application practices and the manufacturer's instructions on mixing and applying them.

Horizontal Cracks

As discussed in Chapter 6, horizontal cracking is often referred to as edgelifting or curling cracks, these appear just at or slightly below the grout-to-concrete interface, usually from ¼ to ¾ in. below the grout into the concrete interface (Figure 7.2). In some cases, this type of cracking is considered not as serious as vertical cracks because it rarely migrates under the machinery. As it gets closer to the equipment

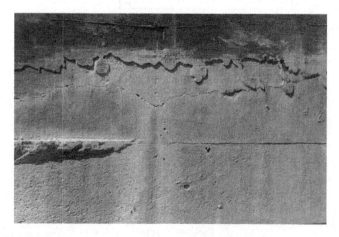

Figure 7.2 Edgelifting or curling cracks appear just at or slightly below the grout-to-concrete interface.

base, the grout bond to the concrete will be maintained because the grout is held in compression by the weight of the machinery and the preload of the anchor bolts. Edgelifting still causes concern because it gives the appearance that the whole grout cap has delaminated. Usually this type of crack will appear if the grout cap extends 6 in. wider than the machinery base. Edgelifting is more likely to occur in a large area where the epoxy grout has been placed as a seal for the top of the foundation or for cosmetic reasons. In such cases, without seeing the crack on the vertical face, you can ascertain the extent of actual separation of the grout cap by tapping on the horizontal surface from the outer edge toward the machinery base.

Preventing horizontal cracking is much easier, and less costly, than repairing it. Some preventative measures follow:

1. Install peripheral dowels that will act as anchors and give the grout something to grab onto (Figure 7.3).
2. Chip off the outer edge of the concrete foundation at a 45° angle to expose rebar (Figure 7.4).
3. Limit the width of the grout shoulder to no more than twice the depth of the grout cap. In some instances, this can pose a problem when it comes to installing the grout due to insufficient access between the form and the equipment (Figure 7.5).

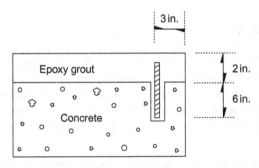

Figure 7.3 Dowels to prevent edgelifting.

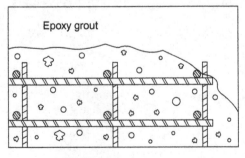

45° Champher to expose
Peripheral rebar

Figure 7.4 Exposing rebar.

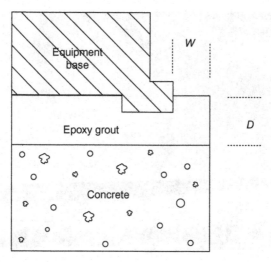

Figure 7.5 Limit the width of the grout.

If the horizontal cracking is only a hairline crack along the vertical face without severe separation, sealing the vertical cracks and doing nothing about the horizontal cracks on the vertical faces of the foundation is usually acceptable. However, if separation is severe, you should remove that section and repour that area. If the area is accessible, a saw with a masonry blade could be used to cut around the bad area to assist in removing the loose grout cap. Clean up the concrete and repour that area for a sound repair. In most cases, expansion joints should be installed at the saw cut or cold joint location or to break up a long mass of epoxy grout that should have had expansion joints in it in the first place.

If the separated cap is to be left in place, drill ⅜–½ in. holes vertically through the grout cap with masonry drills and pump or pour epoxy injection grout into the separated area.

Depending on the area to be filled, you can use a handheld grease gun with the grease fitting attachment removed and a rubber bottle stopper as a seal. Fit it into the drilled holes and pump liquid epoxy into the crack.

Another way is to drill holes on 12–14 in. centers around the separated grout cap and use a copper tubing standpipe with a funnel in which to pour the epoxy. Start in the center and work your way to the outer edges. As you approach the edge, look for signs of the epoxy liquid at the horizontal crack. If possible, use duct tape along the horizontal crack to hold the epoxy liquid until it sets.

> **CAUTION:** After the epoxy starts to flow freely from the horizontal crack, you may not be able to get the duct tape to stick. The tape should be in place prior to the pumping of the liquid epoxy, with vents along the top of the tape.

Figure 7.6 Cracking due to imbedded steel soleplate in epoxy grout.

All of these methods, and others, have been used for years to repair both vertical and horizontal cracks. When properly done, the results have been tight repairs that have extended the life of the grout and the machinery it supports.

Figure 7.6 shows cracking resulting from stress concentration due to steel embedment such as a soleplate with a sharp corner encapsulated by the grout. Corners should have a minimum of ½ in. radius.

Figure 7.7 shows cracking resulting from grout being poured around the unwrapped anchor bolt.

It is possible to formulate an epoxy grout that will not crack. However, such a product does not currently meet the standards required for a machinery grout. Rigid, creep-resistant materials are required to maintain long-term precision alignment. Too little modulus can cause the material to creep under sustained loading. Such dimensional changes could affect the alignment of critical rotating and reciprocating equipment and other precision-aligned machinery. When it comes to precision machinery grouting, it is best to have a more rigid grout with its better support properties, even if the potential for cracking is there. Cracks can be repaired, but a grout that compresses to such an extent that alignment is lost will have to be replaced sooner or later.

CONCLUSION: Designing out stress concentration points is less costly than repairing designed in stress concentration points.

Figure 7.7 Cracking due to unwrapped anchor bolt.

8 Making Deep Pours with Epoxy Grout

Deep, reconstructing pours of epoxy grout were first used by the owners of large pipeline compressors during regrouts. These repairs usually required the removal of 18–24 in. of oil-soaked or damaged concrete. This removed material was replaced with concrete and required between 14 and 28 days of cure. After the concrete was cured, the surface was chipped to remove laitenance and provide a good surface profile for the grout to bond. Using this technology, the length of time a machine was out of service ranged from 6 to 14 weeks and usually resulted in a cold joint at the new concrete to old concrete interface.

To eliminate these long downtimes and designed-in failure points, the pipeline industry started using epoxy grout with pea gravel filler or river rocks to form a rapid-set deep pour compound. This combination significantly reduced machine downtime and eliminated the cold joint problem. As an extra benefit, these pioneers found that by using epoxy grout as a reconstruction material, they significantly reduced the machinery vibrations.

The Grouting Handbook. DOI: http://dx.doi.org/10.1016/B978-0-12-416585-4.00008-0

You may want to rebuild concrete foundations using epoxy grout in some cases because of the advantage of the rapid cure strength (5,000–6,000 psi in the initial 8–10 h of cure). Epoxy grouts have been used for years in making deep foundation capping repairs and regrouting heavy equipment due to the cost savings achieved through the reduction in "out of service time" on critical equipment.

Recent technology and improved grouting materials have resulted in a number of grouting or regrouting methods that have proved successful in a wide range of applications. Several methods or combination of methods, depending on the degree of existing grout and foundation deterioration, can be used successfully in repairing the foundation and regrouting equipment.

Grout and Concrete Removal

Chip the foundation being rebuilt using specialized pneumatic equipment to remove all oil-soaked concrete. Chip the foundation down to clean, sound concrete, removing all horizontal reinforcing bars (Figure 8.1).

Any vertical reinforcing bar damaged during chipping should be replaced and additional vertical reinforcing bar installed on 12 in. centers, if required (also shown in Figure 8.1). This is done to reinforce corners and edges of foundations to reduce or transfer corner stress and to reduce the possibility of edgelifting. Drill holes 1 in. larger than the rebar diameter (and a minimum of 4 in. deep) and grout in dowels using epoxy grout. Refer to the section entitled "Edgelifting: Cause and Cure" in Chapter 6, for more details.

Figure 8.1 Concrete chipped and rebar installed to prevent edgelifting.

Grout Placement

For deep foundation leveling or capping pours of 18 in. or greater, the use of grout aggregate preconditioned to 70°F is highly recommended. Epoxy grout is normally poured to within 6 in. of the equipment base. If the equipment is to be chocked and set later, the grout is poured to within 2–4 in. of the final elevation or up to the soleplates as shown in Figure 8.2.

After the deep foundation leveling pour is made, the final grout pour can be made after the first pour has cured and returned to ambient temperature. Depending on the size of the equipment, ambient temperature, and the amount of grout placed, you may have to allow the grout to cure longer than 24 h and then sandblast or chip the surface before making the final grout pour. A standard unit of epoxy grout should be prepared for grouting in the soleplates, chocks, and equipment base or for making the final pour to the machine base.

Expansion joints may be installed in the foundation capping or regrout pour to reduce the potential for stress cracking due to thermal changes. The basic function or purpose of the expansion or control joints is to reduce the possibility of stress cracks developing in the epoxy grout. The phenomenon or mechanism of stress cracking of aggregate-epoxy resin grout is nonuniform and unpredictable. On large pours, even with the addition of expansion joints between each anchor bolt or every few feet, hairline cracks can develop due to nonuniform stresses caused by temperature extremes or other variables.

Figure 8.2 Grout is poured around but not over soleplates.

What Is the Possibility of Cracking?

In general, experience has shown that cracks on large foundation capping pours can best be controlled by preconditioning the epoxy grout to control the exotherm and employing expansion joints. Temperature extremes must be avoided. Steps to control temperature extremes should be followed when grouting in hot or cold weather. See the sections on hot and cold weather grouting in Chapter 4.

Is Making Deep Pours with Epoxy Grout Too Costly?

It depends on what you are trying to achieve. To complicate and confuse the issue, some advocate rebuilding foundations with *polymer modified concrete* (PMC) products as a way to save money by avoiding those deep epoxy grout pours. The problem with this is that the added labor and machine downtime required overcomes the minimal cost savings between the two materials. Some PMC products claim that they are ready to accept epoxy grout after a 24 h cure. They make this sound too easy.

The American Concrete Institute issued a report (ACI 548.3R-09) in 2009 on PMC. This report is about two pages long and promotes PMC for bridge deck overlay or repair surface. This report shows that PMC has improved physical resistance to impact, abrasion, and chemical resistance. There is nothing in this report to indicate that PMC is suitable for dynamically loaded machinery foundations.

Listing the sequence of events in grouting a piece of machinery using a PMC and epoxy grout will show that an additional day of chipping the concrete and getting ready to make the final pour of an epoxy grout on the PMC will add several additional days to the job before the equipment can be set.

After the concrete foundation surface is initially chipped off and cleaned up, the events should go as follows:

	PMC	Epoxy Grout
Day 1	Build forms	Build forms
Day 2	Pour PMC	Pour epoxy grout
Day 3	Chip PMC	Remove forms and set machinery
Day 4	Clean up foundation again	
Day 5	Pour epoxy grout	
Day 6	Remove forms and set machinery	

NOTE: Do not confuse PMC with Portland cement mixtures containing super plasticizers or other add mixtures.

CONCLUSION: Using epoxy grout is a viable method for reconstructing machinery foundations where time is of the essence. When considering making deep reconstruction pours with epoxy grout there could be instances where it will be necessary to add long runs of horizontal rebar. Discuss this with your epoxy grout manufacturer, or someone knowledgeable in this type of foundation reconstruction.

9 Surface Preparation of Concrete and Steel

If you want it to stick, you've got to prepare the surface correctly.

Surface Preparation of Concrete for Grouting

The Concrete

If epoxy grout develops such a good bond, why do we need to chip the concrete?

Common practice has always been to mechanically roughen the concrete surface prior to grouting. This has been done for cementious grouts as well as for epoxy grouts. This is to remove the *laitenance* from the concrete. Laitenance is formed when the concrete goes from a fluid to a plastic state. More simply stated laitenance collects at the surface of the pour as the water moves out of the concrete mix.

The Grouting Handbook. DOI: http://dx.doi.org/10.1016/B978-0-12-416585-4.00009-2

Because the forms encase the sides of the pour, the only place for the mix water to go is up. This water movement through the fresh concrete is called *hydration* and occurs rapidly during the first few hours after the concrete is poured.

As the water moves up, more and more cement fines are carried to the surface where they collect. How much collects is directly proportional to the amount of water the concrete has in it. The "old timers" called this the "cream." Actually, the term *laitenance* is derived from the French language and means milky.

Pouring epoxy grout on a laitenance-rich surface does little to ensure the success of the installation. First, this laitenance area is the weakest portion of the foundation (not in compressive strength, but in tensile strength). The concrete needs tensile strength to provide a solid substrate for the epoxy grout. Second, depending on how much water was added to the concrete prior to its pouring, the surface laitenance can be several inches deep.

How Far Down Do We Need to Chip?

That's easy: until you get good, sound concrete and aggregate. What is defined as *good, sound concrete?*

Epoxy grout manufacturers who are professional enough to be concerned with the strength of the concrete their grouting material will be applied to usually specify a *minimum* of 350 psi tensile strength concrete. This goes back to the fact that even the epoxy grout with the lowest coefficient of thermal expansion (10.8×10^{-6} in./in./°F) expands and contracts almost twice as much as concrete (6.0×10^{-6} in./in./°F). It is imperative that you reach sound concrete to help overcome this differential.

The best advice that can be given is to chip the concrete on a 45–90° horizontal angle until 50% of the exposed large aggregate shears and remains embedded in the concrete as shown in Figure 9.1.

Jackhammers (Figure 9.2)

Using a jackhammer to roughen the concrete is an unacceptable method. A jackhammer is designed to remove concrete by destroying it. They are designed to be used in an upright position to take full advantage of their weight. They weigh so much that it's virtually impossible to use them to properly chip the surface of a foundation. Additionally, when you use them to chip the surface, they create microfractures in the concrete that could lead to a concrete failure later.

Bush Hammers (Figure 9.3)

This tool is made to fit into a handheld chipping gun and is sometimes referred to as a meat tenderizer. We can only assume that this instrument is used for the following reasons:

- It does not require any sharpening.
- It does not require any special skills to use.
- It is fast.
- It gives the appearance of roughing up the concrete foundation by removing, at the most, about ⅛–¼ in. of concrete.

Figure 9.1 Aggregate shown is imbedded in the concrete.

Figure 9.2 Jackhammers are not recommended for concrete surface preparation.

Figure 9.3 Bush hammers should not be used for surface preparation on concrete machinery foundations.

The bush hammer is designed to be used in an upright position and pulverizes the surface of the concrete. This pulverizing action is akin to beating the surface of the concrete with numerous ball peen hammers very rapidly. Besides removing a negligible amount of concrete, this action actually loosens what little aggregate and sand is in the surface of the concrete. Now when the grout is poured, the concrete it bonds to is already so loose or weakened that its chance of disbonding in the near future is a distinct reality.

Bush hammers were first used by individuals who did not fully understand the dynamics of preparing concrete surfaces for grouting. Their primary concern was to save time and money. Bush hammering a foundation requires significantly less time than chipping. It also produces less waste product, other than this, it does nothing to properly prepare a concrete surface for grouting.

CAUTION: Do not use a bush hammer; stay with the chipping gun.

Chemical Concrete Retarders

Believe it or not, some individuals spray a concrete retarder on the fresh concrete surface of a foundation, wait several hours, and then hose off the concrete surface with high-pressure water. They call this the surface preparation for grouting and it's usually done on large project work by contractors trying to save money and man hours. To date, I do not have enough data to know whether this is a good procedure or a bad one. Until further study can be done or more long-term data is gathered on the effect of the retarder on the concrete at the grout interface, I would advise against this procedure.

Chipping Gun (Figure 9.4)

When all is said and done, this is the tool to use. It may require more effort and clean-up time, but it will produce the best results. These tools are handheld and are easily aimed in any direction necessary. When chipping a concrete foundation, you normally hold them on a 45–90° horizontal angle and use a chisel point. The idea

Figure 9.4 Handheld chipping gun for removing weak surface concrete from machinery foundations prior to grouting.

of using this type of apparatus is to remove concrete without seriously damaging it because concrete with high tensile strength is what you are trying to get down to. So far, this is the only sure way to get there.

> **CAUTION:** Use a good quality, adequately sized chipping gun with a chisel point.

The Preparation of a Steel Surface

Minimum surface preparation for steel is an abrasive blast to Commercial Grade SP-6. You must coat the blasted surfaces with a clear epoxy paint or primer within 8 h. Ideally, a rust-inhibitive epoxy primer should be used. The best surface preparation at the job site is to sandblast to white metal and grout within 72 h. Unfortunately, this is not always possible or practical.

How Well Is the Primer Bonded to the Steel Plate or Machine Base?

This is directly related to the surface preparation of the machine base. Prior to application of the epoxy primer, all steel or iron surfaces must be dry, clean, and free of any previous coatings, rust, and surface contamination. Figure 9.5 shows a properly prepared pump baseplate that has been sandblasted to a white metal finish.

Figure 9.5 A white metal blast with a 3–5 mil surface profile gives the grout a good surface to adhere to.

Figure 9.6 Always inspect the item to be grouted before it is set on the foundation.

How Clean Does the Primer Surface Need to Be When the Grout Is Poured Against It?

Up to one year after the initial application, all that is required prior to grouting is to wipe with a solvent that will not leave a residue. This solvent wipe will remove any contaminants that the machine has picked up during transportation from the manufacturer to the job site. Additionally, it will remove the shine or gloss from the epoxy coating. After one year, the epoxy surface should be, at minimum, wire brushed or lightly sanded with an 80-grit paper.

Your best bet is a lead- and chromate-free, rust-inhibitive, two-component epoxy primer specifically designed for use in conjunction with epoxy grout. The primer must be serviceable in severe industrial and chemical environments and possess excellent adhesion to properly prepared iron and steel surfaces. It should provide rust resistance to ferrous substrates after initial application.

Figure 9.6 shows a new pump baseplate that was stored for over a year in a construction lay down yard waiting to be installed. There had been no underside surface preparation at the time of fabrication or the application of an epoxy primer. The rust shown is from the underside of the baseplate. The contractor had no plans to clean up, only set and grout.

> **CONCLUSION:** For the grout to develop its best bond, both concrete and steel surfaces must be properly prepared and cleaned.

10 Pressure Grouting

Read this chapter first, if you don't read and follow the others.

Pressure Grouting Machinery Baseplates to Eliminate Voids

The injection of epoxy resin under machine bases to fill voids has been used for the last 30 years. By using injection points and vent holes to allow the trapped air to be vented, a filler of epoxy is pumped (or, in some cases, gravity-fed) into the void. This liquid epoxy fills the void, becomes hard, and is as strong, if not stronger than the grout below it, thereby supporting the machine base and reducing resonant vibration.

The need to pressure-inject machinery bases stems from three causes:

1. Machinery bases are not properly prepared to be grouted; that is, not sandblasted, dust-free, grease-free, and so on.
2. Cementious grout that will not bond to steel is used as a cost-saving measure and when mixed and installed has a tendency to bleed its mix water under the base, causing voids as shown in Figure 10.1.
3. Overmixing of the liquid portions of the epoxy grout and overblending of the liquids and the aggregate system can cause air to be entrained in the mixture and be released under the baseplate after grout is placed but before it hardens Figure 10.2.

The Grouting Handbook. DOI: http://dx.doi.org/10.1016/B978-0-12-416585-4.00010-9

Figure 10.1 This piece of cement grout clearly shows that water trapped against the underside of the baseplate will cause voids.

Figure 10.2 Mix completely the resin and hardener until you have a clear amber colored liquid. Do not create vortexing and whip air into the mixture.

These three causes could allow a soft foot condition to exist and a resonant frequency vibration to develop. The resulting vibration can result in excessive seal, bearing, and coupling problems. These problems can be avoided easily by employing proper grouting procedures and techniques when installing baseplate-mounted equipment.

In this chapter, we will discuss the techniques necessary to properly pressure-inject a machinery baseplate or pump base when voids do occur because of improper grouting techniques, actual shrinkage of the grout, or in some cases, when there is

Figure 10.3 Gently tap around on the baseplate to locate void areas. Voids will have a hollow sound.

Figure 10.4 Drilling injection and vent holes into the baseplate.

movement of the machinery base. Filling such voids to restore or achieve good baseplate contact can turn a poor grout job into a successful one.

Locating the Voids

Locating voids under a loose baseplate is a rather simple matter. It requires a small hammer (an 8 oz ball peen works well) (Figure 10.3) and some type of marker. By sounding out the baseplate with the hammer, we can easily locate areas that are not bonded or have voids. Use the marker to outline the void areas.

After you determine the extent of a void, drill holes into the cavity (Figure 10.4). A small void may require only one injection point and one vent point, but usually a

Figure 10.5 Shown is a typical "Zerk" (or grease) fitting used when pressure injecting liquid epoxy.

Figure 10.6 This is a typical hand operated "grease" gun used to inject the liquid epoxy.

multihole layout is required, with injection ports at the outer periphery of the void and a vent port in the center. Different layouts may be required for a large void because the distance between holes should not exceed 12–14 in. Drill the holes vertically if you have access from above. When access is restricted, drill holes at an angle or even horizontally depending on the baseplate or machinery configuration.

After drilling the initial hole into the void, determine the depth of the void by measuring the penetration of a stiff wire. You can drill additional holes to confirm the extent of the void. If the depth checks indicate consistent voids of more than ¼ in. in depth, you will need an epoxy with a filler instead of the normal two-part epoxy injection liquid. A three-component high-flow formulation may even be required if the depth can be measured in inches, which, in turn, will require a larger access hole.

The Injection Equipment

For two-part liquid epoxy injection, the holes usually, are drilled and tapped for ⅛ or ¼ in. pipe fittings (Figure 10.5). Both injection and vent holes are tapped so a vent hole can be used also as an injection point during the final stages of injection. Common grease fittings with pipe threads are used as a means of attaching the pumping mechanism, typically a handheld grease gun (Figure 10.6). When a handheld grease gun is used, its life expectancy will be very short. Because of the nature of the epoxy injection material, any delay can result in the material in the gun hardening. If this occurs, no amount of cleaning will restore the gun to normal operation. It is a good practice to use an inexpensive handheld grease gun because it will normally be thrown away during the course of the job.

Large repair projects with numerous injection points can best be handled by a high-volume pumping system (Figure 10.7) rather than handheld grease guns. Regardless of the type of pumping equipment, you should take care to see that

Figure 10.7 High-volume pumping system. Injection should not be greater than 30 psi.

pressure under the machinery base is limited to 6–30 psi to prevent hydraulic deformation or delamination of a securely bonded section of the base if pressure is
applied too rapidly or to an unvented area. In some cases, the ball checks in some
grease fittings may be removed, or the fittings temporarily not installed or very loose
in the vent holes. In any case, the type of injection equipment should be compatible
with the epoxy-cleaning solvents you are using, and you should clean the equipment
frequently.

In the case of a massive injection project, it may be necessary to obtain an air-
operated reciprocating drum pump. If you use this type of pump, you should install a
pressure regulator on the air side of the pump so the pump's stall speed will be sufficient to prevent overpressuring of the machinery base. Reciprocating drum pumps of
this type should have no greater than a 20:1 ratio and should be able to fit in a 5 gal
bucket.

Liquid epoxy injection material will ruin any type of pumping equipment if
allowed to set up inside it. In the case of pressure pots or reciprocating pumps, you
need to periodically flush these items with a solvent designed for the epoxy being
used.

Mixing

You should mix the two-part injection material in small batches commensurate with
the void size. It is not a good idea to try to split or use only a portion of a large unit
of epoxy injection material. If necessary, though, be sure to accurately measure out
the epoxy resin and hardener to ensure the mixture will cure properly. Quart units

Figure 10.8 Injecting the epoxy liquids into the baseplate voids.

are usually preferred not only because of the short working time of the two-component epoxies (usually 30 min maximum), but also because of the small capacity of the injection (grease) gun. Adequate crew size and proper job planning are essential because the injection process on any piece of equipment should be continuous. In the case of multiple interconnected voids, simultaneous injection with more than one gun may be required.

Application (Figure 10.8)

When you use a hand-operated grease-type gun, remove the end cap and spring plunger and hold the grease gun vertically. The helper will maintain a constant level within the grease gun through the open top as the liquid is being pumped. It is important that the gun always have some liquid level above the plunger so air will not be injected under the base.

Start injection at one of the outer points and continue until material comes out of all the open vent holes. In the case of some API 610 pump bases, the pump base is sloped down from the driven end to the pump. When injecting epoxy resin into these bases, start at the low end and work upward. At times, you may feel some resistance as the pumping is started. This usually occurs when a space of only a few thousandths of an inch, but with a large area, is being filled. If this occurs, stop pumping periodically to allow the pressure to subside as the epoxy mixture flows into restricted voids. Under no circumstance should you force the epoxy into the void. To do this could seriously deform or misalign the baseplate. This is a good reason for monitoring the injection pressure or having the stall speed of a reciprocating pump set at 25–30 psi. Alternately start and stop pumping until the void is full and liquid starts to flow from the vent holes. Sometimes, you need to move onto an adjacent injection point if the void area is large. Remember the goal is to fill the void sufficiently so that injection at the peripheral points will cause flow to be seen at the vents. After the void is completely filled, plug all open holes and do not disturb the grease fittings. In the event of slow leakage into an adjacent void or into a foundation crack, additional pumping can be resumed. If leakage continues, allow the grout to set and attempt a second injection either through the original holes or new ones. Experience and common sense are needed under these circumstances. These procedures may sound complicated; however, many loose pump bases and other types of grouted machinery have been satisfactorily repaired by injection of liquid epoxy.

> **WARNING:** Great care should be taken when pressure-injecting epoxy to fill voids under baseplates. A handheld grease gun can easily generate 10,000 psi and deform, delaminate or actually lift the baseplate off the foundation. Also, two-part liquid epoxy injection grouts are designed to fill thin voids of approximately 0.001–0.250 in. Filling a deep void with a large area may require a specialized product and techniques. Filling deep voids of 1–3 in. with liquid epoxy could result in high exothermic heat that could cause baseplate distortion.

11 Pump Grouting

The last 50 years have seen many pump vibration and seal problems corrected by reengineering the bearings, seals, and coupling. This reengineering can amount to substantial costs over the life of the pump and can still results in machinery problems.

It has become obvious that initial pump installations are done with little or no thought to long-term reliability. The engineering that was done had to do with the dynamics of the pump and how it would perform. The foundation and grouting were left to those with little or no education as to the affect their actions would play on the reliability of that pump. Their concern was cost and time for that moment and their technology of grouting stretched back over 30 years. In other words, they did it the same way for years and never changed.

This technology amounted to pouring cement or epoxy grout up to the bottom flange of the baseplate allowing it to harden (not cure) then make the second pour filling the baseplate cavity. Ninety-five percent of all pumps grouted in this manner require pressure injection to eliminate voids under the baseplate.

The Grouting Handbook. DOI: http://dx.doi.org/10.1016/B978-0-12-416585-4.00011-0

What Was Done to Correct the Problem?

A Case History

In 1986, a major chemical plant in Baytown, TX began using a new epoxy grout technology to increase their pump reliability. They did this by performing the full filling, or single-pour technique on their API 610 pump baseplates. This company recognized it had a unique opportunity to utilize the extremely low exothermic reaction of the ®Chockfast Red epoxy machinery grout, and for the first time were able to fully fill the hollow 6–8 in. underside of API pump baseplates. This new pump grouting technology was previously attempted using other epoxy grouts having exothermic reactions in excess of 200°F. The results were severe baseplate distortion due to the excessive heat generated by those grouts exothermic reaction.

Pumps that had a history of being a "BAD ACTOR" were scheduled for regrouting with epoxy grout as soon as possible (ASAP). Additionally, new API pump installations were given the same attention and provided with a full filling of epoxy grouts.

How Long Did it Take to See the Benefit of This Type of Attention to Details?

It was determined early in the program that the installation cost of using a full filling of the baseplate with epoxy grouts was approximately the same as using epoxy to grout the baseplate lower flange, filling the baseplate cavity with cement or epoxy grout and pressure injecting of the baseplate with liquid epoxy several days later. Installed cost was not a significant issue as the higher labor costs and multiple setup/cleanup costs balanced out against the higher cost of using 100% epoxy grout. The real savings was accrued from future substantial reductions in mechanical seal maintenance costs.

What Did This Plant Hope to Achieve?

The goal of this chemical plant was to improve the mechanical performance of seals, couplings, and bearings by improving alignment through advanced grouting techniques. The following data was provided for 600 pumps out of their total pump population of 2,400 pumps:

1983	6 months MTBF
1986	11 months MTBF
1989	29 months MTBF
1991	19 months MTBF

The decline in mean time between failure (MTBF) between 1989 and 1991 was due to easing of pump installation standards by upper management. This was an effort to achieve higher profits by reducing maintenance costs. This actually cost them more, as by 1993 it returned their pump reliability close to a point where they were 10 years earlier. This is yet another example of quality and dependability versus cost.

So ... What Is the Bottom Line?

In the United States and in other industrial countries the average cost of replacing one pair of tandem mechanical seals in 1990 was about $10,000 per API pump in refinery service. A typical API 10-stage pump with an improved life from 8 to 24 months saved the owner $20,000 in 24 months. Over a 10-year period this savings on a single pump becomes $100,000.

In the case of the plant with 2,550 pumps the estimated savings becomes significant. Of the 2,550 pumps 600 had been regrouted using epoxy grouts. The accrued maintenance savings over 10 years can be calculated as follows:

600 pumps $\times$ $100,000 over a 10 year period $=$ $600,000,000

Keep in mind these are 1990 dollars. In 2013, the cost of these same seals is now $20,000–25,000 per pump.

What Standards Has Industry Set for Acceptable Vibration Limits?

The following has been determined by industry experts and pump manufacturers:

Below 0.1 in. per second (IPS), Anything done to reduce vibration levels at this point are fine tuning efforts only and the time required would be better spent elsewhere.

0.1 IPS, Optimum level. Expect good run life of seals, bearings and couplings.

0.11–0.15 IPS, Acceptable level. Expect fair run life of seals, bearings and couplings.

0.2 IPS, Concern. Investigate to see what is happening.

0.3–0.5 IPS, A problem exists do something quick!!!!!

The Different Ways of Grouting API and ANSI Pump Bases

There are at least seven different techniques for grouting API and ANSI pump baseplates. Some use a combination of cement and epoxy grouts. Some are done to ensure a vibration free installation, whereas others are done in an attempt to cut costs. Here are six different ways to grout a pump baseplate.

A. Completely filling the baseplate cavity with Portland cement grout. Normally, you can accomplish this by making two grout pours, sometimes three. The first pour is to the bottom of the baseplate. The second and possibly third are done to fill the baseplate cavity.

> **RESULT:** High grout shrinkage, poor baseplate contact, and excessive vibration will be the end result. This installation is the cheapest method and always requires pressure injection of epoxy to eliminate voids under the baseplate deck caused by *"bleeding"* of the cement grout.

B. Completely filling the baseplate cavity with nonshrink cementious grout. Normally, you can accomplish this by making two grout pours. The first pour is to the bottom of the baseplate. The second is done to fill the baseplate cavity.

> **RESULT:** There is less grout shrinkage, than with a Portland cement mix; however, poor baseplate contact and vibration will still be possible. This installation is slightly more costly than the preceding method, and usually requires pressure injection of epoxy to eliminate voids due to *"bleeding"* under the baseplate deck.

C. Pouring epoxy grout to the bottom of the pump baseplate, allowing it to cure, then using cementious grout to fill the cavity. Normally, you can accomplish this by making two grout pours. The first pour is to the bottom of the baseplate. The second is done to fill the baseplate cavity.

> **RESULT:** This provides a good support for the lower section of the baseplate through the use of epoxy grout. The possibility of poor baseplate deck contact and vibration still exists, and usually requires pressure injection of epoxy to eliminate voids under the baseplate deck due to *"bleeding"* of the cementious product.

D. Pouring epoxy grout to the bottom of the pump baseplate and use cementious grout to fill the major portion of the cavity then pouring a thin cap of epoxy for the final 1–2 in. This is known as *"sandwiching"* and is mistakenly used to reduce material costs. This technique is used by some individuals or contractors who have underestimated or underbid a project and are trying to cut their cost. Material costs are reduced, but at the expense of increased man hours and a longer cure time for the cement grout (14–28 days). This type of installation will provide minimum vibration damping capability.

RESULT: Material costs are reduced slightly (if at all) but at the expense of increased man hours and a longer cure time for the cement grout. This type of installation will provide low to medium vibration damping capability if done successfully. Pressure injection of epoxy to eliminate voids under the baseplate may be necessary.

NOTE: Cementious grouts will not develop a bond to the cured epoxy grout unless an epoxy adhesive is used.

E. Pouring epoxy grout to the bottom of the pump baseplate let it harden then make a second pour of epoxy grout to fill the cavity. This is known as the *"two pour"* epoxy grout method. The first pour is to the bottom of the baseplate; the second is done to completely fill the baseplate cavity after the first pour has hardened and returned to ambient temperature.

RESULT: The material costs are slightly higher, and the expense of increased man hours is still there. This type of installation will provide good vibration damping capability. It may require pressure injection of epoxy to eliminate voids under the baseplate deck due to air entrainment introduced into the epoxy grout due to aggressive mixing prior to placement.

F. With specialized forming epoxy grout is poured to fill the area from the top of the foundation to the underside of the baseplate deck. This is accomplished at one time and is called the *"single-pour"* method (Figure 11.1).

RESULT: This method usually provides excellent vibration damping provided excessive air is not mixed into the grout. Grouting entirely with epoxy grout provides for *rapid cure* and *decreased downtime* when a quick installation or repair is necessary. This type of installation will provide the very best in vibration damping by eliminating cold joints between pours. In some cases, pressure injection of epoxy to eliminate voids under the baseplate deck due to air entrainment introduced into the epoxy grout due to aggressive mixing prior to placement.

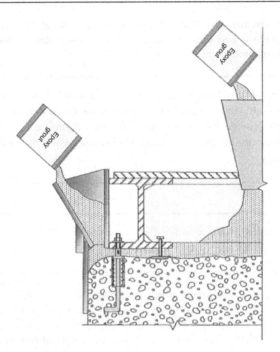

Figure 11.1 Single-pour grouting technique.

NOTE: Baseplate designs vary from flat to sloped deck. Always provide sufficient venting to ensure full contact of the epoxy grout with the baseplate.

G. *Preinstallation grouting* of the pump baseplate.

This method, sometimes referred to as *Inverted Baseplate Grouting*, requires the baseplate to be inverted and the grout poured directly into it to completely fill the baseplate cavity in a single pour (Figure 11.2). Make sure the pedestal supports for the pump and motor ate level and supported prior to the actual grouting. All grout holes and vent holes should be sealed prior to performing this procedure. This type of grouting will be covered in more detail later in this chapter.

RESULT: This method requires no pressure injection to eliminate voids because 100% contact is achieved under the baseplate deck. The only field grouting required is 1–2 in. in the field after you set the baseplate on the foundation and level it.

Figure 11.2 Pregrouted baseplate.

Installation Procedure for Methods A, B, and C Using Cementious Products

- Prepare concrete and baseplate surfaces for grouting. Proper foundation and baseplate surface preparation is essential.
- Wrap the anchor bolts to isolate them from the grout.
- Position and align the baseplate, leaving a maximum 1–1½in. clearance between the bottom of the baseplate and the existing foundation. This clearance can vary depending on the grout manufacturer.
- Install forms around edges of the foundation. Seal the forms to the foundation using putty or silicone caulking. Coat the inside surface of the forms with *paste wax*, or some other type of approved release agent to prevent the grout from bonding to the forms and to facilitate their removal after the grout has cured.
- The top surface of the baseplate must be vented to eliminate air entrapment around the edges of the top surface of the baseplate. If not so equipped vent holes of not less than ½in. in diameter should be installed in the top surface of the baseplate. If the underside of the baseplate is ribbed, you must provide one vent hole in each corner for each individual section defined by the ribbing. On baseplates with no ribbing or other obstructions on the underside, you should install sufficient vent holes to provide sufficient venting.
- Always pour cement grout according to the manufacturers recommendation. Always allow the cement grout to reach its full cure according to the grout manufacturers specifications before proceeding with the final pour of epoxy grout.

Additional Procedures Required for Method D and E, When Making a Cement Grout Fill with an Epoxy Grout Cap

- Prepare concrete and baseplate surfaces for grouting. Proper foundation and baseplate surface preparation is essential.
- Wrap the anchor bolts to isolate them from the epoxy grout and prevent the grout from bonding to them.

- Position, level, and align the baseplate, leaving a maximum 1–1½ in. clearance between the bottom of the baseplate and the existing foundation. This clearance can vary depending on the grout manufacturer.
- Install forms around edges of the foundation. Seal the forms to the foundation using putty or silicone caulking. Coat the inside surface of the forms with *paste wax*, or some other type of approved release agent to prevent the epoxy grout from bonding to the forms and to facilitate their removal after the epoxy grout has cured.
- Make a leveling pour between the foundation and the baseplate anchor bolt flange. Allow the epoxy grout to be poured approximately ¼–½ in. above the bottom surface of the baseplate. Allow the epoxy grout to harden and return to ambient temperature this should take about 8 h, more or less, depending on the ambient temperature.
- Always pour cement grout according to the manufacturers recommendation. Always allow the cement grout to reach its full cure according to the grout manufacturers specifications before proceeding with the final pour of epoxy grout.

Additional Procedures Required for Method F, When Making a Single Pour of Epoxy Grout

- Prepare concrete and baseplate surfaces for grouting.
- Wrap the anchor bolts to isolate them from the epoxy grout.
- Position, level and align the baseplate, leaving a maximum 2 in. clearance between the bottom of the baseplate and the existing foundation.
- Install forms around edges of the foundation and include upper form covers. The forms must be liquid tight and sealed to the foundation using putty or silicone caulking. Coat the inside surface of the forms with paste wax to prevent the epoxy grout from bonding to the forms and to facilitate their removal after the epoxy grout has cured. The upper form covers should be vented to prevent air pockets from forming after the covers are installed.
- The top surface of the baseplate must be vented to eliminate air entrapment around the edges of the top surface of the baseplate. If not so equipped, install vent holes of not less than ½ in. in diameter in the top surface of the baseplate. If the underside of the baseplate is ribbed, you must provide one vent hole in each corner for each individual section defined by the ribbing. On baseplates with no ribbing or other obstructions on the underside, you should install sufficient vent holes to provide sufficient venting.
- A single pour of epoxy grout is made from the top of the foundation completely filling the underside of the pump baseplate. Specialized forming is required in the way of upper form covers. This procedure allows most pumps to be completely grouted within 2 h and provides excellent vibration damping.
- You may need to install a temporary head box (traffic cones make excellent head boxes) at each grout hole in the baseplate to provide a static head for the epoxy grout. As each compartment is filled proceed to the next grout hole and repeat the procedure again until all compartments are filled.
- Sloped-deck baseplates will require some type of waxed cover be installed over each grout fill hole to prevent grout from escaping as the base is filled.
- Allow the grout to reach its full cure before tightening the anchor bolts or installing the pump.

Baseplate size, installation procedures, available clearance, and environmental conditions all play a part in determining which product should be used for a trouble-free installation. Ninety percent of all pump baseplate installations can successfully be performed using an epoxy grout that has a low exotherm and low coefficient of thermal expansion.

CAUTION: Always pour and use grout products, according to the manufacturer's recommendation.

Always follow the grout manufacturers mixing, application, and curing recommendation.

Conventional Epoxy Grouting of Pump Baseplates

In today's modern industrial complexes, the need for equipment reliability is of prime concern to everyone. For years cement grouts were used to install pump bases. Because of their poor bond and shrinkage, it was necessary to pressure inject these bases to eliminate voids. Pressure injecting of baseplates is a time-consuming and expensive repair that may or may not solve the problem.

Because of unacceptable baseplate preparation and poor epoxy grout installation techniques, the high cost and the need to pressure inject pump bases is still with us. The whole concept of grouting is to make the pump base and the foundation monolithic. By doing this, we reduce the natural frequency of the pump base, thus increasing seal and bearing life. Improper grouting techniques can result in repair cost and downtime that could greatly exceed the time and money spent on the initial pump base installation.

The procedures that follow are specifically designed for deep pour, low exotherm epoxy grouts. Deep pour may be classified as an epoxy grout that may be poured to a depth of 12 in. deep in a single pour, thereby allowing for single lift grout pours. The ability to make deep single lift grout pours, coupled with proper foundation and baseplate preparation will reduce the man hours involved in making several lifts, and possibly the need to pressure inject to eliminate voids caused by poor grouting practices.

To begin with, the concrete foundation should be properly cured. It is chipped to provide a good surface profile for the epoxy grout. The foundation must be clean and dry before pouring the epoxy grout. The best way to protect the foundation is to erect a temporary structure over it. This structure will protect the foundation from direct sunlight, which could result in excessive heating and uneven curing of the epoxy grout; also it will allow for environmental control. If the ambient temperature is below 65°F, you need to heat the surrounding area to above 65°F.

To prepare the baseplate, begin by removing the pump, driver, and other accessories mounted to the baseplate. The baseplate should be bare when it is set on the foundation and grouted.

After the equipment is removed, the underside of the baseplate should be inspected. Any additional grout holes, vent holes, or jackscrews should be installed at this time.

After any repairs or modifications are completed, the underside of the pump base should be sandblasted to "white metal." After this step great care should be taken to

Figure 11.3 Using flat plates is not the best method for setting and leveling a baseplate. The grout will not support the baseplate.

prevent any contact with oil, water, or other contaminants that would affect the bond of the grout.

The elapsed time between sandblasting the base and the actual grouting should not allow the surface to "bloom" with surface rust. To prevent this, prime the base-plate underside with a Rust Inhibitive Epoxy Primer, or other approved primer that will create a bond to steel of no less than 1,500 psi and have a dry film thickness of 3 mils.

When setting the baseplate onto the foundation, contractors use several meth-ods to support the baseplate while the grout is being poured and during the curing process (Figure 11.3). Most of these methods usually result in improperly installed baseplates and will result in grout cracking. It is recommended that methods 1–4 *not* be employed when setting pump baseplates.

1. Flat plates cut into squares and stacked one on top of the other until the required elevation of the baseplate is obtained. This technique results in trial and error for proper elevation, and designed in stress risers.
2. Using single or parallel wedges to obtain proper elevation.
3. Incorporating a steel shim pack that is pregrouted in place. (This method is extremely labor-intensive and time-consuming.)
4. Using a nut on the underside of the baseplate to achieve proper elevation (Figure 11.4).
5. Utilizing a jackscrew (Figure 11.5) alongside each anchor bolt is the only sure way to properly set and level pump bases. (This is by far the easiest, most accurate, and least time-consuming method.)

The primary advantage to using a jack bolt is that it can be removed after the grout is cured, therefore allowing the entire pump baseplate to be supported by the grout, not by the leveling devices.

Methods 1–4 do not allow for the grout to accept the load of the baseplate. Furthermore, methods 1, 3, and 4 allow for stress concentration points to be designed

Figure 11.4 Mounting pumps this way shows a distinct lack of understanding about machinery installation.

Figure 11.5 Utilizing jackscrews is the preferred method of baseplate installation.

into the grout. These concentrations points could result in cracking of the epoxy grout at a later time.

Method 4 does not allow for proper tightening of the anchor bolts. Anchor bolts require a minimum of 12-bolt diameters available free length for proper tensioning. This method could result in loose baseplates later, with no way to tighten them short of a regrout.

When using jack bolts, it is recommended that round plate often called a jack pad be used under the jack bolt. This pad can be constructed from ½ in. thick steel plate, old pump shafts or 2 in. diameter rebar. Whichever material is used, it should be a minimum of ½ in. thick and have a minimum diameter of 2 in., or three times the diameter of the jack bolt.

Figure 11.6 Utilizing jack pads for the jackscrew to bear against.

The purpose of this jack pad (Figure 11.6) is to provide a bearing area for the jack bolt and prevent the jack bolt from digging into the concrete during the leveling phase of the pump installation.

There are two ways to mount the jack pad. Some people prefer to secure and level the pad with epoxy putty, while others prefer to simply place the jack pad under the jack bolt and begin the leveling procedure. After the pump base is leveled, the grout forms can be installed.

There are two ways to construct grout forms. Method 1 is to place the forms directly against the foundation. Doing this requires that a seal be placed 1–2 in. below the chipped surface to act as a seal for the epoxy grout. After the grout has cured and the forms removed, then the caulk must be removed, and the interface smoothed by using a specialized epoxy compound.

Method 2 allows the forms to be moved 1–2 in. away from the foundation. Using this technique requires that all foundation surfaces be chipped, and all vertical and horizontal edges be chamfered 2–6 in. to reduce any stress concentration points that may cause cracking in the epoxy grout.

Pouring epoxy grout completely around the pump foundation allows complete encapsulation of the concrete and reduces the possibility of concrete contamination due to oil or product. Also, this method eliminates the need for someone to come back and dress up the foundation.

> **NOTE:** Do not use this method if you would be covering an expansion joint between the foundation and the adjoining pad.

Figure 11.7 Failure to wax forms when using epoxy grout will result in excessive man hours for cleanup.

Whichever method is used, the forms should be constructed of ¾ in. plywood and braced both vertically and horizontally with 2 in. × 4 in. lumber. The face of the forms to come in contact with the epoxy grout should be waxed to prevent bonding of the grout to the forms. Waxing is performed prior to erecting the forms around the foundation. Doing this eliminates the possibility of contaminating the concrete surface. A good hardwood floor paste wax is required. Under no circumstances should liquid wax be used. Apply three coats, allowing the wax to dry before the next coat. Failure to properly wax the grout forms will result in excessive work to remove forms and clean up the grout surface (Figure 11.7).

The grout forms should be liquid tight and sealed to the vertical face with a good caulking material. All inside right angles (90°) should be chamfered to a minimum of 1–2 in. to prevent stress concentration areas and possible cracking of the epoxy grout at a later time.

During the summer, the foundation and equipment to be grouted should be covered with some type of shelter to keep the uncured grout from being exposed to direct sunlight. This covering will also protect the foundation from dew, mist or rain. It should be erected 24 h prior to grouting and remain up until after the grout has completely cured.

In the winter, a suitable covering to allow the foundation and equipment to be completely encapsulated should be constructed. A heating source should be applied

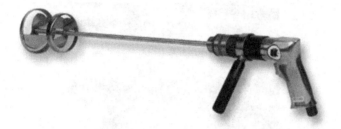

Figure 11.8 Jiffy mixer blade and variable speed drill for mixing liquids.

Figure 11.9 Follow proper mixing techniques to avoid whipping air into the liquids.

so as to raise the foundation and equipment temperature to above 65°F for at least 48 h prior to and after grouting.

Mix the epoxy grout resin and hardener should be mixed in accordance with the manufacturer's instructions for the type of grout being used. Generally, this means mixing the epoxy resin and hardener to a homogeneous state by using a Jiffy type mixer (Figure 11.8) in a slow-speed electric or air drill motor, at a speed of 200-250 rpm. Care should be taken at this point not to whip in air. The mixed epoxy grout resin and hardener (Figure 11.9) should have a clear amber appearance (in cool weather, this could be a milky white color). All parts of the grout (resin, hardener, and aggregate) should have been brought to a temperature of between 65°F and 80°F. This is called preconditioning, and should be accomplished 48 h prior to grouting.

The final mixing and ultimate pouring of the epoxy grout mixture (resin and hardener with aggregate) is accomplished by using a mortar mixer. The liquid is poured into the mixer and the four bags of aggregate are then added. Mixing time will vary

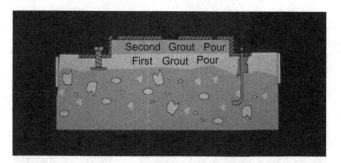

Figure 11.10 This is the conventional way of grouting pump baseplates.

from 2–5 min, depending on ambient temperature, material and foundation temperature. Once the grout is thoroughly mixed, it is then poured or transported via wheelbarrow or buckets to the forms. During the mixing and installation of the epoxy grout, proper safety practice should be employed. Goggles or face shields should be worn by those mixing and pouring the epoxy grout. Protective gloves should be worn by all, and dust masks should be worn by those exposed to the aggregate prior to mixing. Soap and water should be available for periodic hand cleaning should the need arise.

You can install epoxy grout for pump base grouting in two ways. The traditional method is sometimes called *"the two pour method."* (Figure 11.10) This involves pouring epoxy grout only to the bottom flange of the pump base. This pour is allowed to harden, then the remainder of the pump base cavity is poured. The problems that are associated with the two pour method are:

- It takes twice, sometimes three times the man hours to pour a pump base. This is due to equipment cleanup and re-setup to complete the pour.
- The first pour should be allowed to completely cool to ambient temperature before the second pour is made. Failure to do this could result in thermal stress at the interface of the grout which could result in cracking at a later date.

With the single pour method, a set of waxed upper form covers is installed (Figure 11.11), with vent holes drilled about every 12 in. The grout is then poured starting at the pump end and working toward the opposite end. Because most API 610 baseplates are built with a sloped deck, we need to insure that plugs are available for the vent holes, and that covers are available for the grout holes. These plugs and covers should also be waxed to prevent the grout from bonding to them. In some cases, metal plugs are used in the pump base. It may be desirable to allow these plugs to bond to the grout.

Whichever method you employ (the one- or two-pour method), the installer should use some type of head box (Figure 11.12) to ensure complete filling of the pump base. A good head box that can be cut to fit is a typical traffic cone. When grout emerges from the vent hole, a plug should be installed. Grout must flow from each vent hole. After the base is completely filled, the grouting is completed. It is a good practice to have someone stand by with a bucket of grout to add a slight

Figure 11.11 Upper form covers allow baseplates to be completely filled in one pour.

Figure 11.12 Utilizing a traffic cone as a head box for grout installation.

amount to each grout hole as required during the curing process to maintain a head on the grout. Or you can utilize standpipes (Figure 11.13) made from PVC that are removed after the grout hardens.

Cleanup of the uncured epoxy grout is accomplished with soap and water (Figure 11.14). Utilizing a pressure washer also serves to clean spilled grout from concrete surfaces. For cleaning of the baseplate, it is recommended that a good epoxy solvent be used.

After the grout is completely cured, the forms may be removed. Depending on the method used to place the forms (directly against the foundation or 1 in. away), it may be necessary to smooth the vertical face of the foundation with an epoxy faring compound. After this, the foundation may be painted with an epoxy coating.

Figure 11.13 Utilizing PVC standpipes to maintain a constant head pressure on epoxy grout when utilizing a single-pour method on a pump baseplate.

Figure 11.14 Cleaning uncured grout from the mixer is easily accomplished using water.

Pregrouting of API Pump Baseplates

Industry has learned that by proper baseplate grouting, pump vibrations can be significantly reduced and mean runtime between seal, bearing, and coupling failures can be dramatically extended. There have been several excellent articles written about proper pump grouting; however, the problems of pump grouting are still with us.

Many end users specify that a thin-film epoxy coating of some type be applied to the underside of the pump baseplate at the manufacturer's facility. However, the problem stemming from this can result in loose or improperly grouted baseplates. Grout manufacturers are continuously asked, "is your epoxy grout compatible with our primer" or paint? The answer to this is very simple, all epoxy grouts will bond to whatever surface they touch. The question one should be asking is, what is the bond strength of the preapplied primer or paint?

Original equipment manufacturer (OEM's) manufacturing pump baseplates should do everything in their power to assure a good surface profile for the epoxy primer or coating to bond to. Once the baseplate leaves the manufacturing facility, it can be anywhere from 6 months to 2 years before it is actually installed. It arrives at the end user's facility and is either stored in a warehouse or in a laydown yard. From the time it arrives at the job site and goes into storage until the time it is set on its foundation, the surface under the baseplate that the grout will ultimately be required to bond to is usually dirty, oily, or rusted due to rough handling. Very seldom does the installing contractor take the time to inspect the underside of the pump base let alone clean it prior to setting it on the foundation.

Once the baseplate is set and leveled on the foundation, the contractor normally will not begin to install the epoxy grout until there are several pieces of equipment to be grouted. During this waiting period dust and oil can collect as well as rust develop on the underside of the baseplate. When grouting is finally done it is without cleaning the underside of the pump base or removing the pump and its driver for better grouting access. This inattention to detail will definitely result in a poorly bonded pump baseplate and the need to come back and pressure inject to eliminate a soft-foot condition. Problems resulting from pressure injecting to eliminate voids under the pump base by personnel not familiar with this technique can result in hydraulic deformation, or delamination of a securely bonded section of the base if pressure is applied too rapidly or to an unvented area. Care should be taken to see that pressure under the machinery base never exceeds 6–10 psi to prevent these problems.

A new technology has developed over the last 25 years wherein the pump base is inverted and grout poured directly into it at the OEM's shop at the time of initial fabrication (Figure 11.15). This is done after stress relieving and sandblasting but before painting. The grout is allowed to cure prior to the pump base being painted or sent to the machine shop for machining or grinding of the motor pads or pump support pedestals.

What Will the Pregrouting of API 610 Pump Baseplates at the Factory Prior to Machining the Pads Achieve for the OEM?

- Increased rigidity of the pump baseplate will help the OEM meet API 610 nozzle load requirements and reduce test stand vibration so that the assembly will easily meet

Figure 11.15 Size is not a limiting factor when it comes to pregrouting pump baseplates. However, weight may be a consideration when turning the baseplate over.

Figure 11.16 Grout holes and vent holes are not required when pregrouting a baseplate.

the 0.1–0.2 IPS required by API and some end users. ANSI pump specs call for a limit of 0.3 IPS.
- Reduced fabrication costs—no need for grout holes, vent holes, or additional bracing installed in the baseplate (Figure 11.16).
- No need for high volatile organic compounds (VOC) epoxy primers (normally solvent-based and spray-applied).
- Guaranteed 100% grout bond and void-free contact to underside of pump baseplate.
- Maintain sufficient rigidity during transport and lifting to prevent any twisting or bending of the pump base. This could ultimately result in reduced OEM warranty service calls on newly installed pumps incorrectly grouted due to a twisted or deformed baseplate (Figure 11.17).
- Reduce installation problems at the end-user facility by eliminating the need for the general contractor to go through elaborate procedures in the field to achieve a void-free grout job.

Figure 11.17 Increased rigidity offers less chance of deformation during transport.

What is a Contractor Required to Do When Installing a Pregrouted Baseplate?

When installing a pregrouted pump base the contractor is required to do the following:

- Wipe the underside of the pregrouted baseplate with a nonresidue leaving solvent. This is accomplished when the assembly is suspended prior to setting on the foundation.
- Flow approximately 2–3 in. of grout under the pregrouted base (Figure 11.18). Baseplate size will determine clearance. Current procedures in the field require what is commonly known as a two-lift grout pour unless elaborate forming is constructed to allow the pour to be completed in one lift.

This will eliminate any problems that a contractor who is unfamiliar with epoxy grout technology and specialized grouting techniques might have.

From the Pump Owner's Point of View

- The upfront cost associated with the new technology will increase shipping weight from the OEM's facility but not necessarily the freight charges.
- The increased cost of pregrouting should not be any more than what would normally be experienced at the plant level. Actually, the overall cost of grouting from a labor standpoint should be significantly reduced.
- Most OEMs call for the pump and driver to be removed from the base prior to grouting. This allows for the base to be leveled without deformation or distortion from single point support from the jackscrews. Pregrouted pump baseplates will eliminate the need to remove any mechanical components.
- This pregrouting will eliminate the need to pressure inject improperly grouted pump bases. Pressure injection can and will result in serious problems when accomplished by

Figure 11.18 Pregrouted baseplate ready for field grouting. Note the head box incorporated into the form design to enhance flow.

inexperienced personnel. Overpressuring when injecting an epoxy resin system under the baseplate can actually lift or bow the base, and in some cases result in damage to the coupling end section of the pump.

The Benefits of Pregrouting Pump Bases with Epoxy Grout Are

- A bond to the steel baseplate greater than 2,000 psi is achieved.
- A compressive strength greater than 10,000 psi is achieved within 24–48 h after placement.
- 100% bearing area against the baseplate underside.
- Improved machining of the pump baseplate will be achieved due to significantly reduced possibility of tool push off because of increased baseplate rigidity (Figure 11.19).
- The pump base will be easier to grout to the foundation.
- Vibration dampening will be enhanced.
- Baseplate deformation or distortion in the field when using high exothermic epoxy is eliminated.

WARNING: Using an epoxy grout that is designed to be poured on top of concrete or is poured in maximum thicknesses of 6 in. and under should be avoided when using the pregrouted or inverted technique. Because of the high exothermic temperatures that will be generated and the possibility of baseplate deformation that could result because of insufficient heat sink available to the epoxy grout. The use of any epoxy grouting product other than a low exothermic epoxy grout may prove unstable when poured without sufficient heat sink. Consult the grout manufacturer or someone skilled in this type of application.

Figure 11.19 Pregrouted baseplate ready for milling. Pregrouting reduces the possibility of tool push off and tool chatter.

CONCLUSION: MTBF and improved pump reliability can easily be achieved through proper epoxy grouting.

---

12 Skid Grouting

Some new approaches to an old subject.

Conventional Grouting of Skid-Mounted Equipment

Over the years, installation of skid-mounted equipment was done without much thought of long-term effect. If epoxy grout was used, the drawings usually called for a 1–1½ in. maximum grout thickness. No provisions were made for access under the skid, and the installer was usually required to flow the epoxy grout 10–15 ft under the skid through a 1–1½ in. space. To do this, the installer would remove one or two bags of aggregate from the grout to improve its flow, thereby changing the aggregate fill ratio of the grout. This was usually done without the benefit of expansion joints. These practices had several unintended consequences.

The Grouting Handbook. DOI: http://dx.doi.org/10.1016/B978-0-12-416585-4.00012-2

Increased Cost

This is an important factor in any job, but who pays for this? The project engineer has more important things on his mind. The construction superintendent is looking at the overall schedule, so the installer or end user pays, either in dollars or equipment and foundation problems down the road.

Reduced Physical Properties of the Grout

Leaving out aggregate to improve flow is a common practice; however, it is not recommended. The most common reason for removing aggregate to enhance flow is insufficient space (or clearance) between the skid base and the foundation. This lack of clearance comes from

- Improper clearance specified on the installation drawing for the epoxy grout being used.
- Incorrect elevation of the concrete foundation or skid.
- Failure by the installer to remove laitance from the top of the foundation and chip surface properly.

Cracking of the Grout Due to Thermal Stress

The increased exotherm of the epoxy grout from a reduced amount of aggregate depends on the ambient temperature at which the grout is poured. If the ambient temperature is 90°F, the exotherm will be higher than if the grout was poured at 65°F. Again, the removal of aggregate is not recommended.

Grouting skid-mounted equipment for long-term installation can be a difficult process unless some forethought is given to exactly what you are trying to accomplish. The idea behind grouting a skid-mounted piece of equipment is for the grout to form a secure bond to the skid frame and to the concrete foundation to make the entire structure monolithic. A monolithic structure will enable resonate vibrations to be transmitted away from the operating equipment and into the foundation, where they can be dissipated. When you grout this type of equipment with a cementious grout or a Portland cement mix, the installation's ability to transmit these vibrations is greatly reduced.

Skid-mounted equipment originally consisted of a driver and some type of driven compressor or pumping equipment. They were designed as unitized assemblies for temporary service and were installed in the field without benefit of a foundation. Installation usually consisted of dropping them off of the back of a truck and starting them up. The machine was seldom aligned after setup because it was not intended for long-term installation.

Today, skid-mounted equipment is still used and designed as a unitized component or, in some cases, as a complete operating assembly. The majority of installations are long-term and are mounted on concrete foundations.

In the early days, it was concluded that these units would operate with less vibration if the skid frame was grouted to the foundation. This was first done with fluid or flowable cementious grouts. As epoxy grouts became more widely used, they took the place of the cement grouts as the preferred product under skid units with high dynamic operating conditions and loads.

As we discussed in Chapter 1, the general rule of thumb for reciprocating equipment foundation design is for the foundation to be a minimum of five times the mass of the operating equipment. For example, if a reciprocating engine and high-speed compressor assembly weighed 25,000 lb by the rule of thumb, the underlying concrete mass should weigh about 125,000 lb. Sometimes it is not practical or even feasible to install a foundation of this mass for a reciprocating skid-mounted assembly. To compensate for this lack of foundation mass, sometimes the individual skid compartments will be filled with concrete at the time of fabrication. The idea is for this concrete to act as a damping agent by adding mass to the skid. You add the concrete after the skid has been fabricated and painted. Unless you weld some type of reinforcing steel to the inside of the individual skid compartments, the concrete will not perform as expected because it will not bond to the painted steel, and its damping characteristics will be reduced.

Most manufacturers of skid-mounted equipment such as separable compressor packages build their skids to be extremely rigid. These skids are usually decked over with steel plating to restrict oil and water access to the skid compartments. This plating also restricts the access for grouting. A skid-mounted unit can be 14 ft wide and 30 ft long, or larger. Access to both sides of the skid for grout placement is usually available after the skid is placed on the foundation. However, because of its design and the location of the equipment, grouting access to the center of the skid is just about impossible. If access is gained, it is usually because holes are cut in the deck plating so grout can be placed from the center and allowed to flow to both sides. Because the normal clearance between the foundation and the base of the skid is usually less than adequate, usually $1^{1}/_{2}$–2 in. at best (as shown in Chapter 4), the clearance under the skid must be increased to successfully flow epoxy grout over long distances. If clearance cannot be increased, you must use some type of head box to assist the grout in flowing (Figure 12.1).

The major concern with skid grouting is adequate support of the longitudinal and lateral beams of the skid frame. The other concern is getting the grout to flow from one side to the other. Grouting of skid-mounted equipment can be rather difficult at times because the people who write the installation procedures have rarely ever installed one. This is evident by the clearance, or lack of, between the foundation surface and the bottom of the skid.

As the epoxy grout flows across the chipped concrete, it gives up a certain amount of its resin to the dry concrete this forms the bond. As the leading edge of the grout gives up this resin, it becomes more and more rigid, eventually acting as a dam and impeding further progress, or flow, of the grout (Figure 12.2). By using special techniques and designs, you can enhance and assist in the flowing of epoxy grouts over long areas. Access to the underside of the skid is a must. Try these methods to improve access:

- Increase the foundation to skid clearance. This requires more grout.
- Add 6 in. diameter grout pipes to the skid design to allow for placement of the grout from the center of the skid.
- Add access panels in the deck plating when applicable. This is good only if the skid has not been filled with concrete to add mass.

Figure 12.1 Head box can facilitate grout placement under large skid-mounted equipment.

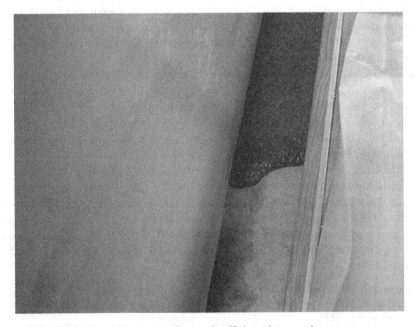

Figure 12.2 The ability of the grout to flow and sufficient clearance is necessary.

Figure 12.3 Expansion joint locations.

Expansion Joints for Skids (Figure 12.3)

There are those who advocate installing expansion joints in the grout under skid-mounted equipment and those who do not. Pouring epoxy grout in large areas without eventually developing a crack somewhere is virtually impossible. Those who support installing expansion joints usually install them so that they will run under the lateral support beams of the skid. As we discussed in Chapter 3, expansion joints bridged by a steel member become almost useless. The possibility of developing a crack within a few inches of an expansion joint bridged by steel is highly likely. Another difficulty is pouring the elastomeric epoxy material to form the primary seal for the expansion joint. After the grout is poured, pouring the primary seal outside of the skid is easy, but pouring it under the skid is virtually impossible.

The most logical procedure is to install no expansion joints; instead, let the grout decide where it feels an expansion joint should be. After the grout cracks, use a small grinder to cut a **V** out of the crack and fill with an elastomeric epoxy material.

You can install expansion joints successfully in skid-grouting systems, providing that the skid units are chocked.

Is the Epoxy Grout Compatible with the Paint on the Skid?

Next to expansion joint use, this seems to be the most common question asked about skid grouting. Most people are concerned about the compatibility of the epoxy grout with the paint or primer applied to the area of the skid that comes in contact with the epoxy grout. The question they should be asking is, "How well is the paint or primer bonded to the

Figure 12.4 Have enough people at the job site to accomplish the various tasks associated with grout placement.

steel?" Epoxy grout will develop some type of a bond to whatever it touches. How well it bonds depends on the surface profile and the extent of contamination. Most epoxy grouts will develop a bond of about 2,000 psi to properly prepared steel. Properly prepared steel is a surface that has been sandblasted to white metal with a 3–5 mil surface profile.

The strength of bond a paint or primer will develop is best determined by asking the manufacturer of that product.

Manpower Requirements for Mixing and Pouring Epoxy Grout

The biggest mistake you can make when trying to reconstruct a large foundation or grout a piece of equipment with epoxy grout is to not have enough workers on hand. Assume, for example, you are pouring 50 units (80 ft³) of epoxy grout and want to accomplish this in less than 2 h. Each unit of epoxy grout will consist of four 47 lb bags of aggregate and approximately 20 lb of liquid. You will probably need two people to handle the aggregate (Figure 12.4).

Two people should be devoted strictly to the mixing of the liquid epoxy and the liquid hardener (Figure 12.5). One should open the liquids and set them up for the other worker to actually mix the liquid components with a Jiffy® mixer. As he completes the mixing of the liquids, he will set the container aside for the aggregate mixers. It is important that the person mixing the liquid does not get too far ahead of or fall behind the people mixing the aggregate. Proper mixing of the liquids should take no more than 3 min per container.

There should be a team of three people who have the task of adding the mixed liquid and the aggregate together in a mortar mixer and blending this into the epoxy grout that will be poured under the machine (Figure 12.6). One will pour the mixed

Figure 12.5 Mixing of the liquids takes at least two people, maybe more to keep up with the mixer team.

Figure 12.6 Mixing teams should work together during grout mixing and placement.

Figure 12.7 Grout placement may require more wheelbarrows than mixers since it may take longer to place than to mix.

liquids into the mortar mixer and ensure proper mixing. The other workers should be responsible for opening the aggregate bags and adding the correct amount of aggregate to the liquid mixture. Depending on the size of the area to be grouted this could require several mixers.

Two or more people should transport the completed mixture of epoxy grout from the mortar mixer to the various areas around the piece of equipment to be grouted (Figure 12.7).

Two or sometimes even three people are needed to ensure that the material is properly placed and spread out, if necessary.

One person should be in charge to oversee the work, coordinate the different stages, and direct the placement of the epoxy grout.

At least nine people or more are required to do the job right. Crew size is dependent on skid size and the amount of grout to be poured.

Pour the Grout and Set the Skid (Figure 12.8)

An interesting procedure was developed in West Texas to facilitate and shorten the amount of time required to grout a skid. Earlier, this chapter discussed the normal 1½ in. clearance usually specified between the skid and the concrete and the cost associated with increasing that clearance.

Figure 12.8 Pouring the grout and setting the skid will probably require some planning but is possible.

Several large skids have been set directly from the truck transporting them into uncured epoxy grout. With a little planning and forethought, you can use the following methods to do the same:

1. Before the truck transport and lifting crane arrives at the job site, chip and form the foundation and set the jackscrew pads in epoxy putty, leveling them in two directions. After the putty has set up, glue 2 in. urethane foam pipe insulation to the jack pads.
2. After the skid arrives at the job site and before the crane lifts it off the truck, back out the jackscrews, install the jam nuts, and reinstall the jackscrews. Set the skid directly on the foundation, level it, and place at the correct elevation using the jackscrews. After the leveling is complete, use the jam nuts to secure the jackscrews.
3. Lift the skid off of the foundation and move it away. Pour epoxy grout over the foundation to a depth of ¼ in. below the final required elevation of the grout. The urethane pipe insulation prevents the grout from getting between the jackscrews and the jackscrew pads, and the jackscrew from coming in contact with the epoxy grout.
4. Set the skid back onto the foundation and into the epoxy grout. After the grout cures, back off the jackscrews, tighten the anchor bolts, and remove the forms.

The result is 100% contact under the skid and a much quicker grout time.

NOTE: Ambient temperature and epoxy grout pot life play an important part in a successful grout installation of this type. Also, having more than one mortar mixer (as a backup) is highly recommended.

Figure 12.9 Grout cap is poured with the pump removed.

Skid Chocking

Chocking of skid-mounted equipment is dependent on the rigidity of the skid itself. Always check with the original equipment manufacturer (OEM) or the skid manufacturer to determine the suitability or practicality of chocking skids.

You can do skid chocking in the following ways:

Base grout cap with epoxy chocks.

Prior to putting the skid on the foundation, install and secure the expansion joints. Fill all anchor bolt sleeves (where required) with a suitable isolation material. Wrap anchor bolts and then pour the epoxy grout a minimum of 2 in. thick. This puts the top of the grout cap 2–2½ in. below the base of the skid's final elevation (Figure 12.9).

After the grout is allowed to cure, the skid may be placed in position and brought to final elevation. After the skid is level and at its final elevation, it is chocked at each anchor bolt with an epoxy chocking compound (Figure 12.10).

• Pouring chocks directly on concrete.

Very similar to pouring chocks on epoxy grout, this technique is possible, but not recommended as the chocks will bear directly on the concrete. If using this type of installation, follow these guidelines:

1. Paint the entire foundation with two coats of epoxy paint per the manufacturer's recommendation.
2. Set the skid. (Jackscrew pads should be round and have no sharp corners.)
3. Level the skid and install wrapping around anchor bolts and jackscrews.
4. Install chock forms.

Figure 12.10 Set the pump and connect it to the piping prior to pouring chocks.

Figure 12.11 Skid-mounted engine driven compressor unit.

Chocking Equipment to the Skid

Euro Gas Systems in Romania use a different approach that works for them. Their technique is to use angle iron forms and pour an epoxy chocking compound under the soleplates of their compressor units. (Figures 12.11–12.16 outline this procedure. Chapter 13 contains a worksheet for calculating epoxy chock size.

Figure 12.12 The epoxy chock is sized and the angle iron forms are welded to the skid around the anchor bolt. Soleplate jackscrews and the anchor are wrapped with pipe insulation to isolate them and prevent bonding of the chocking compound.

Figure 12.13 They are careful to allow enough free length to the anchor bolt to gain maximum clamping force.

Figure 12.14 The soleplates are set in position and rough leveled.

Figure 12.15 The compressor is set and aligned to the prime mover.

Figure 12.16 Once alignment is obtained the epoxy chocking compound is installed to secure the soleplate.

CONCLUSION: Grouting of skid-mounted equipment is not difficult if a little thought is given to what you are doing. Environmental conditions and sufficient manpower play a major role in this. Additionally, sufficient clearance and a head box should be considered.

13 Chocking

Pourable shims: an easier way to set reciprocating compressors and other types of critical equipment.

Epoxy Chocking

Epoxy resins were developed in the mid 1930s, but their use by the industry did not become commonplace until around 1955. The marine industry was the first to use epoxy resin chock to set reciprocating machinery (they did so in the early 1960s). This installation has been followed by more than 125,000 other installations of engines and compressors, both marine and land-based, totaling more than 100,000,000 horsepower, 40,000 gear boxes, and countless other pieces of machinery. Despite the widespread acceptance of epoxy chocks in the marine and the pipeline industries, relatively few engineers fully understand the principles of epoxy resin chocking.

Unfortunately, a major misconception that has inhibited the full use of epoxy chock technology by industries still exists: most engineers assume that iron or steel, being "stronger," must be a more suitable chocking material than the epoxy resin chock. It stands to reason that a high compressive strength material like steel is certainly not a disadvantage in a machinery support system. However, an unnecessarily

The Grouting Handbook. DOI: http://dx.doi.org/10.1016/B978-0-12-416585-4.00013-4

Figure 13.1 Steel chocks were the accepted method for setting large pieces of equipment. This has changed significantly with today's technology for machinery installation.

high one may be accompanied by other undesirable properties that present a distinct disadvantage. One of these is that in order to achieve 100% contact between two steel surfaces, you would need to lap those surfaces to within two light bands, as is done with mechanical seal faces. Steel is handicapped by its high compressive modulus of 30,000,000 psi. A typical steel chock touches only three high spots on one face and possibly only one on the other. The high modulus permits negligible stress distribution. The points of contact are likely to be excessively loaded and susceptible to fretting. Engine designers typically do not concede this because they are in the business of selling support systems made of steel. To date, only a few OEMs have installed one of their new machines on epoxy chocks. Normally, these installations are done with steel soleplate and steel or composite chocks supplied by the OEM or a manufacturer (Figure 13.1).

The large number of anchor bolts featured in the typical industrial engine have tension requirements totaling many times the unit weight. This demonstrates an awareness of some potential problem at the steel chock interface. Epoxy resin chocks require a bolt tension-to-weight ratio of 2.5:1. The conclusion is that the engine is clearly, if unwittingly, designed to accommodate chocks of unsuitable material that do not fit properly. It should be noted that the majority of the engine OEMs readily accept the use of epoxy or composite chocks as substitutes for their steel ones to eliminate costly remachining of the engine bases on older installations.

Steel chocks are designed to fit on existing steel soleplates or rails. Steel chocks are basically U-shaped so they can be removed when the machine must be realigned. The steel chocks are removed, reshimmed, or reinstalled. If the equipment base, soleplate, or rail is worn from fretting, you must address this to achieve acceptable contact.

Because of this U design, steel chocks are free to move, and frequently do when the anchor bolts lose tension. This movement is quite evident by the "hammer"

Figure 13.2 Steel chocks coupled with improper anchor bolt tensioning resulted in loose equipment.

Figure 13.3 Epoxy resin chocks are poured around the anchor and can achieve 98% contact with the machine base.

marks on the steel chocks, showing where they were precision-fitted under the machine by the mechanic (Figure 13.2).

Epoxy resin chocks are cast in place around the anchor bolt and fit perfectly with essentially 100% contact (Figure 13.3). Their compressive modulus is ideal because it will distribute stress caused by foundation or bedplate distortion in service but without significant vertical deflection, so the correct fit and alignment are maintained.

The anchor bolts hold the machine down, whereas the epoxy chocks hold it up. Epoxy chocks work best in conjunction with the proper clamping force of the anchor bolts to maintain the vertical alignment and to prevent horizontal movement of the engine. (Movement is induced by the generated internal and external forces of the operating machine.)

The coefficient of friction between the chock and its mating surfaces is extremely important. The coefficient of friction of an epoxy chock is about three times higher than the coefficient of friction of a steel chock. After dimensional stability, the most important factor in chocking is the available frictional force, so resin chocks have an obvious superiority.

A properly designed epoxy chock, with its high coefficient of friction and correct fit, coupled with sufficient anchor bolt clamping force result in a dynamically stable system. The combined effects of thermal expansion, unbalanced forces, couples, torque reaction, thrust, and any other forces on or within the engine are less than the holding power of the chocks, so no movement takes place.

Some epoxy chock manufacturers advocated the use of a release agent when installing epoxy chocks for two reasons:

1. It makes removing the epoxy chock easier if realignment of the machine becomes necessary.
2. Some people still find it difficult to accept that a reciprocating engine or compressor will grow or expand thermally over its chocks; however, this has been proven. The stresses induced in the chock by the thermal growth of the engine can be significant and become more pronounced with frame length.

Thermal Expansion

Using a coefficient of thermal expansion of 6.0×10^{-6} for a 20 ft engine frame, we see that with a Δt (differential temperature) of 40°F, axial thermal growth of about 60 mils occurs.

A result of this is that thermal growth forces on the engine frame base will be distributed among all chocks. This growth will be transmitted by chock friction. Because the engine is rigid, all the chocks will carry some part of it. Epoxy resin chocks will not slip—metal ones may, but both will deflect in shear to some small extent. The shear angle of metal chocks is considered zero for practical purposes. Epoxy resins have a lower shear modulus than iron or steel, so the shear angle will still be very small. This misconception about the function and working stresses of epoxy chocks has been the major hindrance to a general adoption of resin chocks in new construction.

In the case of some skid-mounted equipment, the precision machining of the equipment base may be eliminated because the chocks will conform to the existing surface profile. The time required to chock with resin chocks can be predicted accurately and is relatively short. Epoxy chocking is usually accomplished in 1 day with the anchor bolt being tensioned in as soon as the chock returns to ambient temperature. In repair work, the potential for time and cost savings is even greater.

In some cases, such as a reciprocating machine installed with steel or cast iron chocks, realignment and rechocking may be required several times in its life. This is due to the fretting of the metal chock, the engine base, and the soleplate or rails. In this case, remachining these items is a major expense and will not solve this problem permanently. Epoxy resin chocks are the obvious solution because their installation does not require the expensive and lengthy process of machining. To this date, no case of epoxy resin chocks wearing out has ever been recorded, which is remarkable when it is considered that many of the engines rechocked between 1968 and 1978 had histories of chronic chock and alignment problems.

The epoxy resin chocks used for machinery chocking also has numerous other applications. Most of these applications are not as demanding as reciprocating engine and compressor chocking. The epoxy resin chock is a good choice because it offers a better, faster, more accurate, and more cost-effective way to install machinery. Unfortunately, epoxy resin chocks are still not attractive to engineering and construction (E and C) firms. In addition to the physical properties already described earlier, another important one is the viscoelastic behavior of the resin. This enables it to withstand very high transient and shock stresses without damage. Several years ago, the US Navy performed high-impact shock tests on epoxy resin chocks of 220 g and observed no damage to the chock. For comparison, a collision of two ships at full speed results in about 1.5 g shock.

All machinery chocking for land-based installation and offshore platforms is carried out in basically the same way:

1. Calculate the epoxy chock size and anchor bolt tension (see worksheet at the end of this chapter).
2. After the engine or compressor is set on the foundation or skid and aligned and the crankshaft web deflections or alignment are satisfactory, clean the area to be chocked with an aromatic solvent that will not leave a residue. Have a drawing available to show position and relative size of the epoxy chocks.

3. After cleaning, start the first stage of damming. This phase uses open cell foam damming about 1 in. thick × ¾ in. taller than the gap between the engine base and the foundation. The overall length of this foam damming should be about 3 ft. This allows sufficient length to facilitate handling and insertion under the engine frame.

4. After the foam damming is in place, spray a mold release or bond breaker into each chock area. After this is done, install the front dam.
5. The front dam is normally made from an angle iron high enough to permit a minimum of ½in. resin overpour and head. Coat the surface of the angle iron dam that will come in contact with the chocking material with a thin coating of high-temperature grease.

6. After the damming is completed, mix and pour the chocking resin into the chock area through the ½–¾in. overpour area.

The chocking material will harden in about 1–2h at 70°. After 18h, remove the forms, back off the jack screws, and tighten the anchor bolts to design specifications. After you tension the anchor bolts, grind the ½in. overpour off with an electric grinder (Figure 13.4).

Machinery such as a diesel generator set that is assembled on a skid or common base must have the alignment between the engine and generator maintained accurately, but this is controlled primarily by the chocking between each component and the skid. The chocking between the skid and the deck (offshore) is not required to maintain an accurate, absolute elevation because the complete generator set has to match up with

Figure 13.4 Grinding off the chock overpour.

only cables, fuel lines, air lines, exhaust pipes, and so on, all of which have a degree of flexibility. Classification approval is generally not required for these applications and a static chock load of 1,200 psi maximum is customary. The chocks are normally sized so the hold-down bolts may be tightened up without exceeding 1,200 psi.

General machinery used on offshore oil rigs and production platforms, such as pumps and winches, can be chocked using the same rules as the generator example. The rules are in fact guidelines set by the epoxy chock manufacturer, so exceptions can be made when necessary and advisable.

For reciprocating machinery, the combined equipment deadweight and anchor bolt clamping force on the chock area is normally designed for 500 psi at a maximum temperature of 176°F.

In the 40-year history of epoxy resin chocking, some of the many thousands of applications have been overloaded, overheated, or otherwise misused. These installations have, however, remained serviceable. The safety margins inherent in the design parameters used are clearly adequate; yet, the applications are practical and economical. Epoxy resin chocking is the optimum method for most reciprocating machinery installations or installations requiring some type of shimming. It is also faster and more cost-effective than other methods—a definite plus. Industry as a whole can benefit immensely by using the epoxy chocking technology.

In the case of reciprocating equipment, the distinct advantage to chocking is that it elevates the machine several inches off the foundation and allows for the free flow of air under the hot running equipment. This flow of air reduces *thermal humping* and misalignment. Before the use of chocks, reciprocating engines were grouted in with the *full-bed* technique (Figure 13.5). The full-bed technique consisted of epoxy or cement grout being poured under and around the machine base encapsulating it. This was all done with the machine at ambient temperature. After the machine was placed in service and reached operating temperature, its thermal growth was restricted by the grout.

Figure 13.5 Grout crack from thermal growth or unwrapped anchor bolt.

Figure 13.6 Adjustable composite chock.

Adjustable Composite Chocks

The use of composite chock (Figure 13.6) is widely used for dynamic heavy machinery installation and along with the epoxy chock has almost replaced the steel chock. Composite chocks offers a precast, ready to use chocking system with most of the same advantages of the pourable epoxy chocks plus the added advantage of being adjustable for hot alignment situations. A two-piece, 2 in., precision-ground split composite chock with 0.035 in. S.S. shims, sits on a hot-rolled steel soleplate with two jack bolts for easy installation and support while grouting.

Horizontal movement resistance, due to the high friction coefficient between the composite chock and metal (frame above and soleplate below), allows equipment to operate smoother. The material used in the composite chock acts as an insulator to greatly reduce heat buildup from the engine frame or other machinery to the foundation. Composite chocks will not fret the bottom of the machinery base or frame. Adjustable two-piece chocks allow shimming up or down, to compensate for "hot alignment" problems.

Two-inch-thick shimmed composite chocks provide free flow of air to minimize effects of "thermal humping" on crankshafts and reduce heat 35–50°F on steel soleplates embedded in epoxy grout. These chocks are machined to ±0.001 in. tolerances and held by bolts through the chock halves, they allow fast alignment corrections if necessary. Their unique dual-purpose level-lift bolts assist in relieving machinery weight on the two-piece chocks for easy insertion or removal of shims. Composite chocks and soleplate Figure 13.7 shows the slot and an inverted view of the soleplate. Note the radiuses on all imbedded corners to relieve stress concentrations in the grout and the welded locking devices to enhance stability in the grout. The slot provides for easy installation. This design provides greater resistance to horizontal forces than conventional steel chocks. This type of chock offers excellent vibration damping and superior chemical resistance.

Figure 13.7 Inverted view of properly designed soleplate.

	Physical Properties of Composite Chocks	
	75°F	266°F
Compressive strength (min.) ASTM D-695	30,000 psi.	20,000 psi.
Shear strength (min.) ASTM D-732	14,000 psi.	
Tensile strength (min.) ASTM D-638	12,000 psi.	
Temperature service range	−40°F to 350°F	

Chocks aid in reducing heat buildup in concrete foundations, which reduces the possibility of their thermal distortion. They also reduce frame distortion commonly experienced by large reciprocating engines installed with full-bed grout.

Dr. Anthony J. Smalley has written several excellent papers on frame distortion and alignment of reciprocating compressors:

1985, "Crankshaft Stress Reduction through Improved Alignment Practices," American Gas Association Project PR15-174.

1987, "Reciprocating Compressor Frame Distortion during a Cold Start," Presented to the "Energy-Sources Technology Conference," Dallas, TX. (ASME Document 87-ICE-4).

1989, "Measurement, Evaluation, and Control of Compressor Alignment," Presented to the American Gas Association Conference, in New Orleans, LA.

One of several conclusions drawn from the research on the two pipeline engines that developed these papers was that the chock-mounted engine showed significantly less transient frame distortion than the full-bed grouted engine.

Additional advantages to using chocking technology are:

- Epoxy grout can be applied to the foundation before the machinery is set. This provides for a quicker installation.

- The use of steel rails and soleplates can be eliminated and a more cost-effective machinery installation accomplished with an epoxy chock.
- A single A36 hot-rolled steel soleplate with the top surface ground to within ± 0.001 of an inch and having a dimension of 10 in. $\times$ 13 in. $\times$ 2 in. thick will cost approximately \$350 and weigh 78 lb.
- The A36 hot-rolled steel chock will be 9 in. $\times$ 8 in. $\times$ 2 in. thick, weighs 33 lb, and costs approximately \$250. This is a total of 125 lb of steel (78 lb of which are imbedded in the epoxy) that will have a different coefficient of thermal expansion than the epoxy grout.
- The approximate total cost of a steel soleplate and one piece steel chock is about \$600.
- The approximate cost of a steel soleplate and composite chock is about \$700.
- The material cost of an epoxy chock is approximately \$1.05 per cubic inch, so for an epoxy chock of comparable size the cost is about \$140.

Because it takes one chock per anchor bolt for any machine, it doesn't take long to see which one is more cost-effective. But cost is not always the determining factor. You must look at many other factors to select a chocking system that will work for you.

What will best suit the reliability and service of the machine? A system with a high coefficient of friction will best serve large reciprocating equipment that is dynamic and has high unbalanced forces.

The following chart is given for comparison of poured-in-place epoxy chocks, Single or two-piece shimable steel chocks, and two-piece shimable composite chocks.

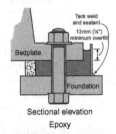

Sectional elevation
Epoxy

Steel

Composite

	Poured-in-Place Epoxy Chocks	Steel Soleplate and Steel Chock	Composite Chock
Requires a soleplate	No	Yes	Yes
Can be installed directly on epoxy grout	Yes	No	No
Ease of installation	Good	Good	Good
Resistance to heat transfer	Excellent	Poor	Excellent
Resistance to fretting	Excellent	Poor	Excellent
High service temperature	Poor	Excellent	Excellent
	(175°F)	+250°F	350°F
Mating with parallel surfaces	Excellent	Good	Good
Compatibility with pitted or fretted machine base	Excellent	Poor	Fair
Ability to make quick alignment changes	Poor	Excellent	Excellent
Coefficient of friction	Excellent	Poor	Excellent
	0.07	0.23	0.07

Epoxy Chock Design Worksheet

1. __ lb (unit deadweight); ÷__ (no. anchor bolts) =__ lb (deadweight per bolt)
2. __ lb (dead wt. per bolt) +__ lb (bolt load from table) =__ lb (total load per bolt)
3. __ lb (total load per bolt); ÷__ psi (chock load) =__ in.^{2} (chock area under frame to the next highest whole number)
4. __ in.^{2} (chock area under frame); ÷__ in. (chock width under frame) =__ in. (chock length)
5. __ in. (chock width under frame from number four) + 0.75 (overpour) =__ (chock length from number four) ×__ in. ×__ (chock depth (thickness)) =__ in.^{3} (total volume per chock)
6. __ in.^{3} (total volume per chock) ×__ (no. of chocks) =__ in.^{3} (chocking compound required)
7. __ in.^{3} (chocking compound required) × 1.2 (loss/waste);__ in.^{3} (volume/unit) =__ (units needed to the next highest whole number)

Epoxy Chock Configuration
Dimensions

Chock *length* is right to left as you look at it.
Chock *width* is front to back as you look at it.
Chock *depth* is top to bottom as you look at it.

14 Grouting Specifications

If you want it done correctly, write it out correctly, step by step. The maximum you can expect is the minimum you will accept.

Machinery Grout Specifications

Why Do Most Original Equipment Manufacturer Equipment Installation Specifications Call for Only 1–2 in. of Epoxy Grout?

Most epoxy grouting specifications used by original equipment manufacturers (OEMs), refining, petrochemical, and industry in general, are modified cement

The Grouting Handbook. DOI: http://dx.doi.org/10.1016/B978-0-12-416585-4.00014-6

grouting specifications written more than 30 years ago. These specifications call for a thin pour of a cementious grout. Thirty years ago, the industry didn't know what we know today about the advantages of using deep-pour epoxy grout for reconstructing damaged foundations and reducing downtime. Most of the specifications used today are technically incorrect and grossly out of date. They have been added to and modified so many times they create more problems than they solve.

Vibration and premature failure of machine components usually stem from a poor installation, system design, foundation design, type of grout, installation *procedures*, or anchoring system, just to name a few. On many occasions, I have been asked to look at a grout installation that was perceived as having a problem. Is the problem with the grout itself, or with the initial engineering or construction of the installation? My experience has been that all grouts are satisfactory when used in accordance with their design and application parameters. The quality of the grout installation is greatly dependent on the skill, knowledge, and experience of those designing and performing it. All too often, I hear stories like the following:

- Engineering experience: The design engineer has 15 years in designing machinery foundations (In 15 years, this person has never left the engineering and construction (E and C) company office).
- Construction experience: "We've always done it this way." (Does that make it right?) If a project comes in under budget and on time, you can usually be sure that shortcuts were taken somewhere.
- Inspection experience: "I have 10 years of experience in following new construction and machinery installation." (Actually, he had only 2 weeks of experience recycled 518 times).
- Grout installer experience: "I've been doing this for years." Actions and performance indicate he actually had 4 months experience recycled 60–90 times.

Unfortunately, during the construction and installation phase of a project, the reaction to any problem is, "What did we do last time?" The correct response should be, "What action would be most effective and provide the best machine reliability in this case?"

Most of the problems encountered in the field spring from the fact that our industrial-based knowledge system supports the concept that repetition is the only way to learn.

The OEM Grouting Specifications

These are published by equipment manufacturers who really do not want to get involved in any component below the bottom of their equipment. Some of their installation specifications simply read, "Grout with an acceptable product," or they issue a grouting specification that is extremely generic, contradicts itself, and is technically incorrect with today's grouting technology. Most OEMs will not specify or recommend a particular epoxy grout even though serious warranty issues may arise from the use of an inappropriate epoxy grout (one with an extremely high exotherm, high coefficient of thermal expansion, or poor installation techniques suggested by the manufacturer).

The E and C Firm Specifications

Most E and C firms do not have individual grouting specifications for specific types of equipment. They have a generic grouting specification that was written years ago with the help of a grout manufacturer (usually a salesman) for a product that no longer is manufactured or has been reformulated. This specification is usually modified to reflect certain aspects of a particular installation. Rewriting grouting specifications is not a priority of most E and C because they tend to use their customers' specifications. Most E and C grouting specs list several grouting products as approved. These products have been reviewed by someone within the firm and are deemed acceptable for use. Immediately after this listing is the catch phrase "or approved equal." This opens the door for any general-purpose grout or the cheapest to be used. All epoxy grouts have different physical properties.

End-User Specifications

The end user is the ultimate user of the machine. The machine will be set and grouted at a location within their plant. They have a long-term interest in how this machine will perform and how it is installed. Unfortunately, their grouting specs are not much better than the people who are selling the machine or doing the engineering. Their specs are usually written by someone with a limited amount of epoxy grout technology. Also, updating these specifications is not a priority within the company.

Material Specifications and Procedures for Epoxy Grouting or Chocking of Reciprocating Engines and Compressors

General Guidelines

- This specification covers epoxy grouting of reciprocating equipment on concrete foundations using rails, soleplates, full-bed grouting, or epoxy chocks.
- Prior to any work being performed, you should contact the grout manufacturer or his representative and arrange a prejob meeting to discuss all aspects of equipment grouting. The contractor should be present at this meeting. If you are not using an outside contractor, the plant maintenance foreman or crew supervisor should be in attendance.
- The machinery engineer should define the responsibilities of the grout manufacturer or his representative and will direct to whom the grout manufacturer or his representative will report during the course of the project or job.
- Distribute a written summary of this meeting to all parties concerned prior to the start of the job.

Material and Testing Requirements

- Epoxy grout should meet the following minimum requirements:
- Minimum compressive strength—12,000 psi as per ASTM C-579-modified Method B.

- Epoxy grout creep shall be less than 0.005 in./in. when tested in accordance with ASTM C-1181 method.
- Epoxy grout linear shrinkage shall be less than 0.080% and thermal expansion less than 17×10^{-6} in./in./°F when tested in accordance with ASTM C-531.
- Pot life—2–3 h at 72°F.
- Cleanup solvent—water.
- Even aggregate distribution throughout the cured grout with no resin-rich surface.
- Maximum coefficient of thermal expansion below 24.0×10^{-6} (ASTM D-696).

Material Storage

- Store all grout materials in a dry area in original unopened containers.
- Precondition all epoxy grout components to a minimum of 65°F and a maximum of 80°F for at least 48 h prior to mixing and placement.

Preparation of Foundation: New Concrete

- Perform shrinkage test as per ASTM C 157 on new concrete; determine when shrinkage is complete.
- If no shrinkage test is performed, approximate cure time as follows:
  Standard cement, 21–28 days *minimum* (five-bag mix)
  High early cement, 7 days *minimum* (six- to seven-bag mix)
- Concrete compressive strength shall be a minimum of 4,500 psi and have a minimum tensile strength of 450 psi as per ASTM C 496.

Prepare the concrete surface for new or old concrete as follows:

- The concrete foundation should be dry and free of oil.
- Chip the concrete to expose a minimum of 50% broken aggregate and, when applicable, remove all laitenance and provide a rough surface for bonding.

> **NOTE:** Use hand chipping guns only. No jackhammers or bush hammers will be permitted.

- After chipping, free the exposed surfaces of dust and concrete chips by sweeping, vacuuming, or blowing using an oil- and water-free compressed air supply from an approved source.
- After you have chipped and cleaned the foundation, protect it to prevent it from becoming wet or contaminated. If grouting will not be accomplished within a reasonable amount of time, coat the chipped surface with a clear epoxy coating to prevent contamination of the surface and to provide a surface that can be easily cleaned should the need arise prior to grouting.
- Examine foundation bolts for damaged threads and take corrective action. Protect the foundation bolt threads during the equipment leveling and grouting operations. Always allow a minimum of 12 times the bolt diameter for free stretch. Accomplish this by wrapping the bolt threads with weather stripping, duct tape, polyurethane pipe insulation, or other approved materials.

- If the bolts are sleeved, fill the sleeves with expanding urethane foam to within 112 in. of the top of the sleeve. After the foam has solidified, apply a cap of an elastomeric epoxy to prevent contaminants from migrating down the anchor bolt later. This also prevents the annular space around the bolt from being filled with grout.

Jackscrew Leveling Pads

- Set and prepare jackscrew leveling pads as follows:
  - Use pads made of 3 in. diameter; ½ in. thick 4140 or similar type round stock if available.
  - Sandblast pads white metal and prime with a thin film (3–5 mil dry film thickness) epoxy coating.
  - Radius pads on the edges to reduce stress concentrations in the grout.
  - Do not use square leveling pads.
  - When applicable, use a high compressive strength epoxy putty to install the pads, which provides a 100% bearing area surface. When you use this procedure, the pads will be leveled. Consult the grout manufacturer or machinery engineer as to when this procedure should be used.
  - Grease or wrap the jackscrews, when used, with duct tape to facilitate their removal after the grout has cured.

Preparation of Engine or Compressor Base, Rails, or Soleplates

- If you are not going to grout immediately, paint the base, rail, or soleplates with one to two coats of thin-film epoxy coating to give a dry film thickness of 3 mil. Fully cure this coating prior to placing the grout.
- Sandblast surfaces of the base, rail, or soleplates that will come in contact with the epoxy grout to a white metal finish.
- Radius vertical and horizontal edges of the baseplate, rail, or soleplate that come in contact with the epoxy grout to a minimum of ½ in. to reduce stress concentrations in the grout.
- If you do not grout the epoxy-coated base, rails, or soleplates within 30 days, rough the coated surface up with a wire brush to remove the bloom or shine. Remove all dust produced by brushing. Clean these surfaces and dry them prior to placing the grout.

Forming

- Coat all forming material coming in contact with the grout with three coats of a good quality hardwood floor paste wax. Do not use liquid wax.
- Take care to prevent any wax from contacting the concrete foundation or the base, rail, or soleplate.
- Make forms liquid tight to prevent the leaking of grout material. Seal cracks and openings with a good quality silicone sealant.
- Eliminate all inside right angles by using ½–2 in. chamfer strips. The machinery engineer or the grout manufacturer must be consulted when in doubt.

Expansion Joints

- Install expansion joints, when used, at locations called out on the installation drawings, as directed by the machinery engineer, by the grout manufacturer, or by his representative.

- Make expansion joints from 1 in.-thick Styrofoam or redwood. Discuss variations with the machinery engineer or the grout manufacturer.
- Incorporate the *secondary seal* design into expansion joints in which the bottom of the expansion joint comes in contact with the foundation.
- To seal the bottom of the expansion joint, mix an elastomeric epoxy with a minimum elongation factor of 200% at 0°F with #3 grit dry blasting sand at approximately four to seven parts sand to one part elastomeric epoxy to form a nonslump mortar consistency. Layer the mix 1–2 in. thick by 3 in. wide on top of the concrete where the expansion joint is to be installed. Set the expansion joint into the mix and press down. When cured, this mixture will form a secondary seal to prevent any contaminants from reaching the concrete.
- Make provisions to allow for removal (after the grout has been poured and cured) of ½–1 in. of the exposed expansion joint surface. Fill this area with the elastomeric epoxy without sand.
- In the area where you are going to use the elastomeric epoxy, all surfaces must be free of any contaminants that would prevent the material from bonding.
- When deemed necessary by the equipment engineer, incorporate horizontal rebar into the grout design so that it will not penetrate an expansion joint.

Mixing

- Prior to mixing and pouring the epoxy grout, the machinery engineer, the grout manufacturer, or their representative shall inspect the area to be grouted for the following:
  - Base, rail, soleplate, and concrete cleanliness.
  - Chamfer strips installed and forms waxed.
  - Foundation bolts properly wrapped and sealed.
  - Expansion joints properly prepared and sealed.
  - Mixing equipment clean and suitable.
  - Ambient and material temperatures within limits.
  - Record ambient temperatures at the beginning of mixing and at the completion of pour and give them to the machinery engineer who will record the data in the permanent equipment records.
  - Make sure that the foundation temperature is a minimum of 60°F.
  - Make sure that mixing equipment is free of all foreign material, moisture, and oil and that the equipment is in good working order and properly sized. Mix three-component epoxy grout materials in a mortar mixer at 15–20 rpm.
  - All personnel handling or working with the grouting materials should follow safety instructions as directed by the equipment engineer.
  - Use only full units of epoxy resin, hardener, and aggregate in preparing the grout.
  - Blend the epoxy resin and hardener for 3–4 min with a properly sized Jiffy® mixer and a ½ in. drill motor, at a speed of no greater than 200–250 rpm.
  - Add the aggregate immediately after completing the liquid blending to fully wet the aggregate. Follow the directions of the grout manufacturer or his approved representative.

Placement

- When required, prepare a suitable head box to hydraulically force the grout under the engine/compressor base or rail.
- Continually pour the grouting until the placement of epoxy grout is complete under all sections of the engine/compressor base, rail, or soleplate. Pour grout from one side, corner, or end to prevent air entrapment.

- Do not use mechanical vibrators to place the grout under the engine/compressor base, rail, or soleplate.
- If required by the equipment engineer, make one 2 in. × 2 in. × 2 in. test cube from each batch number of grout placed. Tag the sample(s) with the equipment number on which the batch was used and where in the foundation the batch was placed.
- Consult the grout manufacturer if testing is required.

Finishing

- If a cosmetic appearance is required or desired, contact the grout manufacturer for directions pertaining to the specific grout system you are using.
- Leave forms in place until the grout has cured. The surface of the grout should be firm and not tacky to the touch. Contact the grout manufacturer for the appropriate cure time based on ambient temperature.
- Where required, dress all edges of the epoxy grout smooth by grinding.

Epoxy Chocking

- Inspect the grout cap, rail, or soleplate for a smooth, clean, oil-free surface.
- Make sure that the machine base is clean and smooth. Fill all pitted surfaces with a high-bond epoxy-fairing compound.
- Size epoxy chocks as follows: Size chocks under the mainframe for a load of 500 psi. Consult the grout manufacturer for proper chock sizing.
- Wrap all foundation bolts with ¼ in. weather stripping to prevent the epoxy from coming in contact with them. If required, fill the area where the foundation bolt penetrates the engine or compressor base with duct seal.
- Wrap leveling or jackscrews that will be in the chock area with duct tape to facilitate their removal after the chock has cured.
- Install open cell foam rubber dams. The height of the foam dam will be ½ in. greater than the chock thickness.
- After you install the foam rubber dams but before you install the front chock dams, spray the chock area with epoxy release agent. Do this under the guidance of the chock manufacturer or his representative.
- Install the front chock dams. These dams should be made of angle iron and must provide a minimum ½ in. head above the engine or compressor base surface. Allow ¾ in. clearance between the angle iron and the engine or compressor base. Spray the area of the angle iron exposed to the epoxy or coat it with a release agent.
- Seal the bottom of the angle iron with a good quality silicone sealant.
- Do not pour the epoxy chock until the chock manufacturer or his representative has approved the preceding installation.

Cleanup

- Immediately after you complete the grouting or chocking, clean all tools and mixing equipment using water or an approved solvent.
- Dispose of all unused mixed epoxy materials and cleanup residue in accordance with instructions from the facility environmental engineer or local authority.
- Direct any questions concerning these specifications to the machinery engineer, the grout manufacturer, or his direct representative.

Material Specifications and Procedures for Epoxy Grouting of API and ANSI Pump Baseplates

General Guidelines

- This specification covers epoxy grouting of mechanical equipment on concrete foundations using baseplates, rails, or soleplates.
- Contact the grout manufacturer or his representative prior to performing any work and arrange a prejob meeting discuss all aspects of equipment grouting. Make sure that the contractor is present at this meeting. If you are not using an outside contractor, make sure that the plant maintenance foreman and/or crew supervisor are in attendance.
- The machinery engineer shall define the responsibilities of the grout manufacturer or his representative and will direct to whom the grout manufacturer or his representative will report during the course of the project or job.
- Distribute a written summary of this meeting to all parties concerned prior to starting the job.

Material Standards and Applicable Tests

- Epoxy grout should meet the following minimum requirements:
- Fire resistant as per ASTM D-635.
- Minimum compressive strength—12,000 psi (ASTM C-579, Method B).
- Pot life of 2–3 h at 72°F.
- Cleanup solvent—water.
- Grout must have low enough exotherm to provide deep-pour capability up to 18 in. deep × 7 ft × 7 ft or greater.
- Even aggregate distribution throughout the cured grout with no resin-rich surface.
- Maximum coefficient of thermal expansion 11.2×10^{-6}, ASTM D-696.
- Grout aggregate of the low dust type.

Material Storage

- Store all grout materials in a dry area in original unopened containers.
- Precondition all epoxy grout components to a minimum of 65°F and a maximum of 80°F for at least 48 h prior to mixing and placement.

Preparation of Foundation: New Concrete

- Perform a shrinkage test as per ASTM C 157-80 on new concrete to determine when shrinkage is complete.
- If no shrinkage test is performed, approximate the cure time as follows:
  Standard cement 21–28 days minimum (five-bag mix)
  Hi-early cement 7 days minimum (six- to seven-bag mix)
- Make sure that the concrete compressive strength is a minimum of 3,500 psi.
- Make sure that the concrete tensile strength is a minimum of 350 psi as per ASTM C 496-90.

Concrete Surface Preparation for Old and New Concrete

- Make sure that the concrete foundation is dry and free of oil.
- Chip the concrete to expose a minimum of 50% aggregate to remove all laitenance and provide a rough surface for bonding. Install dowels to prevent edgelifting or peripheral rebar or expose dowels on new concrete at this time.
- After chipping, sweep, vacuum, or otherwise blow the exposed surfaces free of dust and concrete chips using oil- and water-free compressed air from an approved source.
- After the foundation has been chipped and cleaned, protect it from becoming wet or contaminated. If you cannot finish grouting within a reasonable amount of time, coat the chipped surface with a clear epoxy coating to prevent contamination of the surface and to provide a surface that can easily be cleaned should the need arise prior to grouting.
- Examine foundation bolts for damaged threads and take corrective action. Protect the foundation bolt threads during the equipment leveling and grouting operations. If practical, allow a minimum of 12 times the bolt diameter for free stretch. Do this by wrapping the bolt threads with weather stripping or other approved materials.
- If the bolts are sleeved, fill the sleeves with elastomeric material or expanding urethane foam to prevent the annular space around the bolt from being filled with epoxy grout.

Jackscrew Leveling Pads

- Set and prepare jackscrew leveling pads as follows:
- Make sure that pads are made of 3 in. diameter, ½ in. thick 4140 steel or similar type round stock material, if available.
- Sandblast pads to white metal and prime with an epoxy coating.
- Radius pads on the edges to reduce stress concentrations in the grout.
- Do not use square leveling pads.
- When applicable, use a high compressive strength epoxy putty to install the pads, which provides a 100% bearing area surface. When you use this procedure, the pads will be leveled. Consult the grout manufacturer or machinery engineer as to when to use this procedure.
- Grease or wrap jackscrews, when used, with duct tape to facilitate their removal after the grout has cured.

Preparation of Baseplate, Rails, or Soleplates

- Radius vertical and horizontal edges of the baseplate, rail, or soleplate that come in contact with the epoxy grout to a minimum of ½ in. to reduce stress concentrations in the grout.
- Sandblast surfaces of the baseplate, rail, or soleplates that will come in contact with the epoxy grout to a white metal finish.
- If you are not going to do the grouting immediately, paint the baseplate, rail, or soleplates with one to two coats of thin-film epoxy coating to give a dry film thickness of 3 mil. Make sure that this coating is fully cured prior to placing the grout.
- If the epoxy-coated baseplate, rails, or soleplates are not grouted within 30 days, roughen the coated surface with a wire brush to remove the bloom or shine. Remove all dust produced by brushing. Clean and dry these surfaces prior to placing grout.
- Before grouting API pump baseplates, remove all mounted equipment and grout only the pump baseplate.

- Install ½ in. diameter vent holes in API pump baseplates to prevent air entrapment in compartments isolated by angle iron or I-beam bracing. Consult the equipment engineer or grout manufacturer or his representative for specific locations.

Forming

- Coat all forming material coming in contact with the grout with three coats of a good quality paste floor wax. Do not use liquid wax.
- Take care to prevent any wax from contacting the concrete foundation or the baseplate.
- Make forms liquid tight to prevent the leaking of grout material. Seal cracks and openings with a good quality silicone sealant.
- Eliminate all inside right angles by using chamfer strips, ½–2 in. Consult the machinery engineer or the grout manufacturer when in doubt.

Expansion Joints

- Install expansion joints, when used, at locations as called out on the installation drawings, as directed by the machinery engineer or by the grout manufacturer.
- Construct expansion joints from 1-in. thick Styrofoam or redwood. Discuss variations with the machinery engineer or the grout manufacturer.
- Incorporate the *secondary seal* design into expansion joints where the bottom of the expansion joint comes in contact with the foundation.
- To seal the bottom of the expansion joint, mix an elastomeric epoxy with a minimum elongation factor of 200% at 0°F with #3 grit dry blasting sand at approximately four to seven parts sand to one part elastomeric epoxy to form a nonslump mortar consistency. Layer the mix 1–2 in. thick by 3 in. wide on top of the concrete where the expansion joint is to be installed. Set the expansion joint into the mix and press down. When cured, this mixture will form a secondary seal to prevent any contaminants from reaching the concrete.
- Make provisions to allow for removal (after the grout has been poured and cured) of ½ in. of the exposed expansion joint surface. Fill this area with the elastomeric epoxy without sand.
- Some pump bases do not conveniently allow placement of expansion joints. In such cases, you can locate the joint under the cross bracing beams using 1 in. Styrofoam or similar compressible material. You usually cannot remove this type of expansion joint after placing the epoxy grouting materials. Therefore, allow for the visible portion of the expansion joint to be removed and seal with an elastomeric. The remaining part of the expansion joint will remain under the cross brace beam, permanently sealed.
- Where the elastomeric epoxy is to be used, all surfaces must be free of any contaminants that would prevent the material from bonding.

Mixing

- Prior to mixing and pouring of the epoxy grout, the machinery engineer, the grout manufacturer, or their representative shall inspect the area to be grouted for the following:
    - Baseplate, rail, soleplate, and concrete cleanliness.
    - Installed chamfer strips and waxed forms.
    - Foundation bolts properly wrapped and sealed.
    - Expansion joints properly prepared and sealed, if applicable.
    - Mixing equipment clean and suitable.

- Ambient and material temperatures within limits.
- Record ambient temperatures at the beginning of mixing and at the completion of the pour and give them to the machinery engineer, who will record the data in the permanent equipment records.
- Make sure that the foundation temperature is a minimum of 60°F.
- Make sure that the mixing equipment is free of all foreign material, moisture, and oil and that the equipment is in good working order and properly sized. Mix three-component epoxy grout materials in a mortar mixer at 15–20 rpm.
- All personnel handling or working with the grouting materials should follow safety instructions as directed by the equipment engineer.
- Use only full units of epoxy resin, hardener, and aggregate in preparing the grout.
- Blend the epoxy resin and the hardener for 3–4 min with a properly sized Jiffy® mixer and a ½ in. drill motor, at a speed of 200–250 rpm.
- Immediately after you complete the liquid blending, add the aggregate and blend to fully wet the aggregate. Do this under the direction of the grout manufacturer or his approved representative.

Placement

- When required, prepare a suitable head box to hydraulically force the grout into the pump baseplate cavities.
- Continually grouting until the placement of epoxy grout is complete under all sections of the rail or compartments of the baseplate. Pour grout from one side, corner, or end to prevent air entrapment.
- Do not use mechanical vibrators to place the grout under the baseplate, rail, or soleplate. Rakes or similar tools may be used to place the grout if necessary.
- If required by the equipment engineer, make one 2 in. × 2 in. × 2 in. test cube from each batch number of grout placed. Tag the sample(s) with the equipment number on which the batch was used and where in the foundation the batch was placed.
- Consult the grout manufacturer if testing is required.

Finishing

- If a cosmetic appearance is required or desired, contact the grout manufacturer for directions pertaining to the specific grout system being used.
- Leave forms in place until the grout has cured. The surface of the grout should be firm and not tacky to the touch. Contact the grout manufacturer for the appropriate cure time based on ambient temperature.
- Sound the top of the pump baseplate for voids. If you locate any, drill two holes in each void at opposite corners of the cavity. Tap both the holes and fit one with a pressure grease fitting. Use the other hole as a vent hole and plug it when you complete the injection. Fill the void with unfilled epoxy grout using a grease gun. Take care to prevent lifting or deforming the baseplate.
- Dress all edges of the epoxy grout where required smooth by grinding.

Cleanup

- Immediately after you finish grouting, clean all tools and mixing equipment using water or an approved solvent.

- Dispose of all unused mixed epoxy materials and cleanup residue in accordance with instructions from the facility environmental engineer or local authority.
- Direct any questions concerning these specifications to the machinery engineer, the grout manufacturer, or their direct representative.

Material Specifications and Procedures for Epoxy Grouting of Baseplates, Rails, or SolePlates

General Guidelines

- This specification covers epoxy grouting of mechanical equipment on concrete foundations using baseplates, rails, or soleplates.
- Contact the grout manufacturer or his representative prior to any work being performed and arrange a prejob meeting to discuss all aspects of equipment grouting. Make sure that the contractor is present at this meeting. If you are not using an outside contractor, make sure that the plant maintenance foreman and/or crew supervisor is in attendance.
- The machinery engineer shall define the responsibilities of the grout manufacturer or his representative and will direct to whom the grout manufacturer or his representative will report during the course of the project or job.
- Distribute a written summary of this meeting to all parties concerned prior to starting the job.

Materials

- Make sure that epoxy grout meets the following minimum requirements:
  - Fire resistant as per ASTM D-635.
  - Minimum compressive strength—12,000 psi (ASTM C-579).
  - Pot life of 2–3 h at 72°F.
  - Cleanup solvent—water.
  - Grout must have low enough exotherm to provide for deep-pour capability.
  - Even aggregate distribution throughout the cured grout with no resin-rich surface.
  - Maximum coefficient of thermal expansion 11.2×10^{-6}, ASTM D-696.
  - The grout aggregate is of the low dust type.

Material Storage

- Store all grout materials in a dry area in original unopened containers.
- Precondition all epoxy grout components to a minimum of 65°F and a maximum of 80°F for at least 48 h prior to mixing and placement.

Preparation of Foundation: New Concrete

- Perform a shrinkage test as per ASTM C 157-80 on new concrete to determine when shrinkage is complete.

If no shrinkage test is performed, approximate the cure time as follows:

Standard cement: (5-bag mix)	21–28	days	minimum
Hi-early cement: (6- to 7-bag mix)	7	days	minimum

- Concrete compressive strength is a minimum of 3,500 psi.
- Concrete tensile strength is a minimum of 350 psi as per ASTM C-496.

Concrete Surface Preparation for Old or New Concrete

- Make sure that the concrete foundation is dry and free of oil.
- Chip the concrete to expose a minimum of 50% aggregate to remove all laitenance and to provide a rough surface for bonding. Install dowels to prevent edgelifting or peripheral rebar or expose dowels on new concrete at this time.
- After chipping, sweep, vacuum, or blow the exposed surfaces free of dust and concrete chips using oil- and water-free compressed air from an approved source.
- After you have chipped and cleaned the foundation, protect it to prevent it from becoming wet or contaminated. If you cannot finish grouting within a reasonable amount of time, coat the chipped surface with a clear epoxy coating to prevent contamination of the surface and to provide a surface that can easily be cleaned should the need arise prior to grouting.
- Examine foundation bolts for damaged threads and take corrective action. Protect the foundation bolt threads during the equipment leveling and grouting operations. If practical, allow a minimum of 12 times the bolt diameter for free stretch. Do this by wrapping the bolt threads with weather stripping or other approved materials.

Jackscrew Leveling Pads

- If the bolts are sleeved, fill the sleeves with elastomeric material or expanding urethane foam to prevent the annular space around the bolt from being filled with epoxy grout.
- Set and prepare jackscrew leveling pads as follows:
  - Make sure that pads are made of 3 in. diameter, ½ in. thick 4140 steel or a similar type of round stock material, if available.
  - Sandblast pads to white metal and prime with an epoxy coating.
  - Radius pads on the edges to reduce stress concentrations in the grout.
  - Do not use square leveling pads.
  - When applicable, use a high compressive strength epoxy putty to install the pads, which provides a 100% bearing area surface. When you use this procedure, the pads will be leveled. Consult the grout manufacturer or machinery engineer as to when this procedure is to be used.
  - Grease or wrap jackscrews, when used, with duct tape to facilitate their removal after the grout has cured.

Preparation of Baseplate, Rails, or Soleplates

- Radius vertical and horizontal edges of the baseplate, rail, or soleplate that come in contact with the epoxy grout to a minimum of ½ in. to reduce stress concentrations in the grout.
- Sandblast surfaces of the baseplate, rail, or soleplates that will come in contact with the epoxy grout to a white metal finish.
- If you cannot do the grouting immediately, paint the baseplate, rail, or soleplates with one to two coats of thin-film epoxy coating to give a dry film thickness of 3 mil. Fully cure this coating prior to placing the grout.

- If you do not grout the epoxy-coated baseplate, rails, or soleplates within 30 days, toughen the coated surface with a wire brush to remove the bloom or shine. Remove all dust produced by brushing. Clean and dry these surfaces prior to placing the grout.
- Install ½ in. diameter vent holes if required in large baseplates to prevent air entrapment. Consult the equipment engineer, grout manufacturer, or his representative for specific locations.

Forming

- Coat all forming material coming in contact with the grout with three coats of a good quality paste floor wax. Do not use liquid wax.
- Take care to prevent any wax from contacting the concrete foundation or the baseplate.
- Make forms liquid tight to prevent leaking of grout material. Seal cracks and openings with a good quality silicone sealant.
- Eliminate all inside right angles by using chamfer strips, ½ – 2 in. The machinery engineer or the grout manufacturer must be consulted when in doubt.

Expansion Joints

- Install expansion joints, when used, at locations as called out on the installation drawings, as directed by the machinery engineer, or as directed by the grout manufacturer.
- Construct expansion joints from 1 in. thick Styrofoam or redwood.
  Discuss variations with the machinery engineer or the grout manufacturer.
- Incorporate the "secondary seal" design into expansion joints in which the bottom of the expansion joint comes in contact with the foundation.
- To seal the bottom of the expansion joint, mix an elastomeric epoxy with a minimum elongation factor of 200% at 0°F with #3 grit dry blasting sand at approximately four to seven parts sand to one part elastomeric epoxy to form a nonslump mortar consistency. Layer the mix 1–2 in. thick by 3 in. wide on top of the concrete where the expansion joint is to be installed. Set the expansion joint into the mix and press down. When cured, this mixture will form a secondary seal to prevent any contaminants from reaching the concrete.
- Make provisions for removal (after the grout has been poured and cured) of ½ in. of the exposed expansion joint surface. Fill this area with the elastomeric epoxy without sand.
- Some baseplates or rails do not conveniently allow placement of expansion joints. Discuss this with the machinery engineer, the grout manufacturer, or their representative.
- In the area where the elastomeric epoxy is to be used, all surfaces must be free of any contaminants that would prevent the material from bonding.

Mixing

- Prior to mixing and pouring of the epoxy grout, the machinery engineer, the grout manufacturer, or their representative should inspect the area to be grouted for the following:
  - Baseplate, rail, soleplate, and concrete cleanliness.
  - Installed chamfer strips and waxed forms.
  - Foundation bolts properly wrapped and sealed.
  - Expansion joints properly prepared and sealed, if applicable.
  - Mixing equipment clean and suitable.

- Ambient and material temperatures within limits:
  - Record ambient temperatures at the beginning of mixing and at the completion of pour and given them to the machinery engineer, who will record the data in the permanent equipment records.
  - Make sure that the foundation temperature shall be a minimum of 60°F.
  - Make sure that the mixing equipment is free of all foreign material, moisture, and oil and that the equipment is in good working order and properly sized. Mix three-component epoxy grout materials in a mortar mixer at 15–20 rpm.
  - All personnel handling or working with the grouting materials should follow safety instructions as directed by the equipment engineer.
  - Use only full units of epoxy resin, hardener, and aggregate in preparing the grout.
  - Blend the epoxy resin and the hardener for 3–4 min with a properly sized Jiffy® mixer and a ½ in. drill motor at a speed of 200–250 rpm.
  - Immediately after completing the liquid blending, add the aggregate and blend to fully wet the aggregate. Do this under the direction of the grout manufacturer or his approved representative.

Placement

- When required, prepare a suitable head box to hydraulically force the grout under the baseplate, rail, or soleplate.
- Continually grout until the placement of epoxy grout is complete under the baseplate, rail, or soleplate. Pour grout from one side only to prevent air entrapment.
- Do not use mechanical vibrators to place the grout under the baseplate, rail, or soleplate. You may use rakes or similar tools to place the grout if necessary.
- If required by the equipment engineer, make one 2 in. × 2 in. × 2 in. test cube from each batch number of grout placed. Tag the sample(s) with the equipment number on which the batch was used and where in the foundation the batch was placed.
- Consult the grout manufacturer if testing is required.

Finishing

- If a cosmetic appearance is required or desired, contact the grout manufacturer for directions pertaining to the specific grout system being used.
- Leave forms in place until the grout has cured. The surface of the grout should be firm and not tacky to the touch. Contact the grout manufacturer for the appropriate cure time based on ambient temperature.
- Sound the top of the baseplate for voids. If you locate any, drill two holes in each void at opposite corners of the cavity. Tap both of the holes and fit one with a pressure grease fitting. Use the other hole as a vent hole and plug it when the injection is completed. Fill the void with unfilled epoxy grout using a grease gun. Take care to prevent lifting or deforming the baseplate.
- Dress all edges of the epoxy grout where required smooth by grinding.

Cleanup

- Immediately after you finish grouting, clean all tools and mixing equipment using water or an approved solvent.

- Dispose of all unused mixed epoxy materials and cleanup residue in accordance with instructions from the facility environmental engineer or local authority.
- Direct any questions concerning these specifications to the machinery engineer, the grout manufacturer, or their direct representative.

Material Specifications and Procedures for Full-Bed Epoxy Grouting for the Skid Section of Mechanical Equipment

Also contained in this specification are the chocking procedures for various components of the skid assembly.

General Guidelines

- This specification covers epoxy grouting of skid-mounted mechanical equipment on concrete foundations for long-term installations.
- Contact the grout manufacturer or his representative prior to performing any work and arrange a prejob meeting to discuss all aspects of the skid grouting. Make sure that the contractor is present at this meeting. If you are not using an outside contractor, make sure that the plant maintenance foreman and/or crew supervisor is in attendance.
- The machinery engineer should define the responsibilities of the grout manufacturer or his representative and will direct to whom the grout manufacturer or his representative will report during the course of the project or job.
- Distribute a written summary of this meeting to all parties concerned prior to staring the job.

Materials

- Make sure that the epoxy grout meets the following minimum requirements:
  - Fire resistant as per ASTM D-635.
  - Minimum compressive strength of 15,000 psi (ASTM C-579 Method B).
  - Pot life of 2–3 h at 72°F.
  - Compatible with using water for cleanup.
  - Have a low enough exotherm to provide for a single lift deep pour up to 18 in. deep × 7 ft × 7 ft or greater.
  - Have even aggregate distribution throughout the cured grout with no resin-rich surface.
  - Have a maximum coefficient of thermal expansion less than 16.0×10^{-6} per degree Fahrenheit, as per ASTM D-696.
  - Use a low dust type aggregate.

Material Storage

- Store all grout materials in a dry area in original unopened containers.
- Precondition all epoxy grout components to a minimum of 65°F and a maximum of 80°F for at least 48 h prior to mixing and placement.

Preparation of Foundation

- Test new concrete as follows:
  - Perform shrinkage test as per ASTM C 157-80 on new concrete to determine when shrinkage is minimal.
  - If no shrinkage test is performed, approximate cure time as follows:
    Standard cement (three- to five-bag mix)   28 days minimum
    Hiearly cement (six- to seven-bag mix)   7 days minimum
  - Make sure that concrete compressive strength is a minimum of 3,500 psi when tested in accordance with ASTM C-39 and C-31.
  - Make sure that concrete tensile strength is a minimum of 350 psi as per ASTM C 496-90.

Concrete Surface Preparation for Old or New Concrete

- Make sure that the concrete foundation is dry and free of oil.
- Chip the concrete to expose a minimum of 50% aggregate to remove all laitenance and provide a rough surface for bonding. Install dowels to prevent edgelifting or peripheral rebar or expose dowels on new concrete at this time.

NOTE: Use only hand chipping guns. Do not use jackhammers or bush hammers.

- After chipping, blow the exposed surfaces free of dust and concrete chips using oil- and water-free compressed air from an approved source. You also may vacuum the concrete surface.
- After you chip and clean the foundation, cover it to prevent it from becoming wet or contaminated.
- Examine foundation bolts for damaged threads and take corrective action. Protect the foundation bolt threads during the equipment setting, leveling, and grouting operations. Always allow a minimum of 12 times the bolt diameter for free stretch. Do this by wrapping the bolt threads with weather stripping or other approved materials.
- If the bolts are sleeved, fill the sleeves with elastomeric material or expanding urethane foam to prevent the annular space around the bolt from being filled with epoxy grout.

NOTE: Under no circumstance shall the epoxy grout be allowed to flow into the anchor bolt sleeves.

Jackscrew Leveling Pads

Set and prepare jackscrew leveling pads as follows:

- Make sure that pads are made of minimum 3 in. diameter, ½ in. thick 4140 steel or a similar type of round stock material, if available.

- Radius pads on the edges to reduce the possibility of stress concentrations being developed in the epoxy grout.
- Do not use square leveling pads.
- When applicable, use a high compressive strength epoxy putty to install the pads, which provides a 100% bearing area surface. When you use this procedure, the pads will be leveled. Consult the grout manufacturer or machinery engineer as to when this procedure should be used.
- Wrap jackscrews, when used, with duct tape to facilitate their removal after the grout has cured.

Preparation of Skid Frame for Grouting

- Radius vertical and horizontal edges of the skid frame base that come in contact with, or are embedded in the epoxy grout, to a minimum of ⊠in. to reduce stress concentrations in the grout.
- Sandblast surfaces of the skid frame that will come in contact with or are embedded in the epoxy grout to a white metal finish.
- If you cannot do the grouting immediately, paint the sandblasted areas to be grouted with one to two coats of thin-film epoxy coating. Make sure that this coating is fully cured prior to placing the grout.
- If you do not grout the epoxy-coated areas within 30 days, roughen the coated surface with a wire brush or sandpaper to remove the bloom or shine. Remove all dust produced by brushing or sanding. Clean and dry these surfaces prior to placing grout.
- Install access holes in the skid frame or the metal decking to provide access to compartments isolated by the I-beam framework of the skid. Consult the equipment engineer, grout manufacturer, or his representative for specific locations.

Forming

- Make sure that forming around the skid is sufficient to allow for adequate placement of the epoxy grout.
- Coat all forming material coming in contact with the grout with three coats of a good quality hardwood floor paste wax. Do not use liquid wax.
- Take care to prevent any wax from contacting the concrete foundation or the skid frame.
- Make forms liquid tight to prevent leaking of epoxy grout material. Seal cracks and openings with a good quality silicone sealant.
- Eliminate all inside right angles by using chamfer strips, 1–2 in. Consult the machinery engineer or the grout manufacturer when in doubt.

Expansion Joint

- Install expansion joints, when used, at locations as called out on the installation drawings, as directed by the machinery engineer, or as directed by the grout manufacturer.
- Construct expansion joints from 1 in. thick Styrofoam or redwood. Discuss variations with the machinery engineer or the grout manufacturer.
- Incorporate the "secondary seal" design into expansion joints where the bottom of the expansion joint comes in contact with the foundation.

- To seal the bottom of the expansion joint with a secondary seal, mix an elastomeric epoxy with a minimum elongation factor of 200% at 0°F with #3 grit dry blasting sand at approximately four to seven parts sand to one part elastomeric epoxy to form a nonslump mortar consistency. Layer the mix 1–2 in. thick by 3 in. wide on top of the concrete where the expansion joint is to be installed. Set the expansion joint into the mix and press down. Then pack the excess around the bottom of the joint material approximately 1 in. high. When cured, this mixture will form a secondary seal to prevent any contaminants from reaching the concrete.
- Make provisions to allow for removal (after the grout has been poured and cured) of ½ in. of the exposed expansion joint surface. Fill this area with the elastomeric epoxy *without sand*.
- Some skid assemblies do not conveniently allow placement of expansion joints. In such cases, you can locate the joint under the lateral cross bracing beams by using ¼ in. plywood, 1 in. Styrofoam or similar compressible material. You usually cannot remove this type of expansion joint after you place the epoxy grouting materials; therefore, you should make an allowance for the visible portion of the expansion joint to be removed and seal it with an elastomeric epoxy. The remaining part of the expansion joint will remain under the cross brace beam, permanently sealed.
- In the area where the elastomeric epoxy is to be used, lightly abrade all surfaces to enhance the bond and make sure that they are free of any contaminants that would prevent the elastomeric epoxy material from bonding.

Mixing

- Prior to mixing and pouring of the epoxy grout, the machinery engineer, the grout manufacturer, or their representative should inspect the area to be grouted for the following:
  - Skid frame and concrete for proper surface preparation and cleanliness.
  - Grout form chamfer strips installed at proper elevation and forms adequately waxed.
  - Foundation bolts properly wrapped and sealed.
  - Expansion joints properly prepared and sealed, if applicable.
  - Mixing equipment clean and suitable.
  - Ambient and material temperatures within limits.
  - Record the ambient temperatures at the beginning of mixing and at the completion of pour and give them to the machinery engineer who will record the data in the permanent equipment records.
  - Foundation temperature should be a minimum of 65°F.
  - Make sure that the mixing equipment is free of all foreign material, moisture, and oil and is in good working order and properly sized (4 ft³ max). Mix three-component epoxy grout materials in a mortar mixer at 15–20 rpm max.
  - All personnel handling or working with the epoxy grouting materials should follow safety instructions as directed by the equipment engineer.
  - Only full units of epoxy resin, hardener, and aggregate should be used in preparing the grout.
  - Blend the epoxy resin and the hardener for 3–4 min with a properly sized Jiffy® mixer and a ½ in. drill motor, at a speed of 200–250 rpm.
  - Immediately after completing the liquid blending, add the low dust aggregate and blended to fully it. Do this under the direction of the grout manufacturer or his approved representative.

Placement

- When required, prepare a suitable head box to hydraulically force the grout under the skid framework.
- Continually grout until the placement of epoxy grout is complete under all individual sections of the skid. When you get approval from the machinery engineer or the grout manufacturer, pour the epoxy grout from both sides of the skid at once. Then visually verify that the epoxy grout completely flows under all support beams.
- Do not use mechanical vibrators to place the grout under the skid assembly.
- If required by the equipment engineer, make one 2 in. × 2 in. × 2 in. test cube from each batch number of grout placed. Tag the sample(s) with the equipment number on which the batch was used and where in the foundation the batch was placed.
- Consult grout manufacturer if testing is required.

Finishing

- If a cosmetic appearance is required or desired, contact the grout manufacturer for directions pertaining to the specific grout system being used.
- Leave forms in place until the grout has cured. The surface of the grout should be firm and not tacky to the touch. Contact the grout manufacturer for the appropriate cure time based on ambient temperature.
- Dress all edges of the epoxy grout where required smooth by grinding.

Cleanup

- Immediately after you finish grouting, clean all tools and mixing equipment using water or an approved solvent.
- Dispose of all unused mixed epoxy materials and cleanup residue in accordance with instructions from the facility environmental engineer or local authority.
- Direct any questions concerning the preceding grouting specifications to the machinery engineer, the grout manufacturer, or his direct representative.

Epoxy Chocking of Skid Assemblies for Short-Term Installation

- Inspect the concrete foundation or the grout cap for a smooth, clean, oil-free surface.
- Clean and smooth the concrete foundation, skid rail, engine, or compressor base. Fill all pitted surfaces with a high-bond epoxy-fairing compound.

Size epoxy chocks as follows:

- Size chocks under the skid as per the epoxy chock manufacturer, his representative, or the machinery engineer.
- Design chocks under the engine or the compressor as per the epoxy chock manufacturer, his representative, or the machinery engineer.
- Wrap all foundation or mounting bolts to be chocked with ¼ in. × 1 in. weather stripping to prevent the epoxy chock from coming in contact with them. If required, fill the area where the foundation or mounting bolt penetrates the skid rail, engine, or compressor base with duct seal to prevent the chocking compound from locking in the bolt.

- Wrap leveling or jackscrews that will be in the chock area with duct tape or other approved methods to facilitate their removal after the chock has cured.
- Install open cell foam rubber dams. The height of the open cell foam dam will be ½in. or greater than the ultimate chock thickness.
- After you install the foam rubber dams but before you install the front chock dams, spray the chock area with epoxy release agent. Do this under the guidance of the chock manufacturer or his representative.
- Installing the front chock dams.
- Front chock dams should be made of angle iron and must provide a minimum ½in. head above the underside of the skid rail base surface. Allow ¾in. clearance between the angle iron and the skid rail base. Coat the area of the angle iron exposed to the epoxy chocking compound lightly with a nonmelt grease.
- Seal the bottom of the angle iron with a good quality silicone sealant.
- Do not pour the epoxy chock until the chock manufacturer or his representative has approved the preceding installation.
- Some epoxy chock manufacturers recommend the use of a release agent to be applied to the underside of the skid base to come in contact with the chocks. This agent is usually sprayed on with an aerosol can. Check with the epoxy chock manufacturer for specific details as to when and how this material should be applied.

Cleanup

- Immediately after you complete the chocking, clean all tools and mixing equipment using an approved solvent.
- Dispose of all unused mixed epoxy materials and cleanup residue in accordance with instructions from the facility environmental engineer or local authority.
- Direct any questions concerning the preceding grouting or chocking specifications to the machinery engineer, the epoxy chock manufacturer, or his direct representative.

General Specifications for Epoxy Injection of Grouted Baseplates

General Guidelines

- This specification covers liquid epoxy injection of baseplates grouted with epoxy or cementious grout.
- Prior to performing any work, contact the grout manufacturer or his representative and arrange a prejob meeting to discuss all aspects of pressure injection. Make sure that the contractor is present at this meeting. If you are not using an outside contractor, make sure that the plant maintenance foreman and/or crew supervisor is in attendance.
- The plant engineer should define the responsibilities of the grout manufacturer or his representative and will direct to whom the grout manufacturer or his representative will report during the course of the project or job.
- If required, distribute a written summary of this meeting to all parties concerned prior to starting the job.

Materials

Epoxy injection material should meet the following minimum requirements:

- Minimum compressive strength 10,000 psi (ASTM C-109).
- Pot life of 45 min at 72°F.
- Linear shrinkage no less than 0.0002 in./in. (ASTM D-2566).
- Tensile strength of 5,000 psi (ASTM D-638).
- All epoxy material should be Type IV, Grade 1, Class C, and long-pot-life, and should be new and manufactured within the shelflife limitations set forth by the manufacturer.
- All epoxies should comply with the requirements of ASTM C 881, standard specification for epoxy-resin-base bonding system for concrete.
- The epoxy system should be a two-part adhesive material containing 100% solids.
- The epoxy used should be insensitive to the presence of moisture.
- The viscosity of the epoxy used for injection work should be low enough (about 400 cps at 77°F) to completely fill clearance areas as small as 10 mil.

Material Storage

- Store all epoxy materials in a dry area in their original unopened containers.
- Precondition all epoxy components (if required) to a minimum of 70°F and a maximum of 80°F for 24 h prior to mixing and placement.

Epoxy Injection

- Inject epoxy using a two-component epoxy system designed for such application.
- Make sure that the injection system can handle injection pressures up to a maximum of 300 psi to ensure complete penetration of the epoxy compound under the baseplate.

Baseplate Preparation

- Prior to preparing the surface of the baseplate, clean the immediate area of the vicinity receiving epoxy injection. Blowout all injection points shall be using compressed air and checked them for port-to-port transmission.
- Drill all injection and vent holes and tape them with national pipe thread taper (NPT) threads to receive a grease fitting or pipe plug.

Method of Injection

- All entry points (and vent ports when plugged) must allow for the injection of epoxy without escaping.
- Entry points should be grease fittings with ball checks. The size of the grease fitting depends on the amount of material to be injected and the size of the area to be injected. Grease fitting size can range from ⅛–½ in.
- Establish vents around the area to be injected so that injection epoxy penetrates the void area completely.
- If you use reciprocating air pumps for injection, make sure that they are a positive displacement type with stall capability to provide a positive pressure control to prevent overpressuring.

Installation

- Inject the epoxy adhesive at the first point through the grease fitting with sufficient pressure to advance the epoxy to the adjacent vent. When clear liquid epoxy issues from the vent without any bubbles, you may shift the injection process to the point by installing a grease fitting into the vent port in which the epoxy is exiting. Continue in this manner of point-to-point injection until you have injected each void under the baseplate or until you have injected the entire length of the baseplate in one continuous operation.
- Report any condition other than normal to the engineer or owner's representative.
- Do not use solvents to thin the epoxy system introduced into the cracks or joints.

Surface Finishing

- Unless required, leave the baseplate surface as is. Removing the grease fittings could provide access for contaminants at a later date.
- If a flush surface of the baseplate is required after you complete the injection process, replace the grease fittings with pipe plugs and grind all pipe plugs until smooth.

Introduction

Surface Finishing

15 Drawings Only

I want to thank ITW Polymer Technologies for providing the following drawings to better illustrate concepts discussed in earlier chapters.

The Grouting Handbook. DOI: http://dx.doi.org/10.1016/B978-0-12-416585-4.00015-8

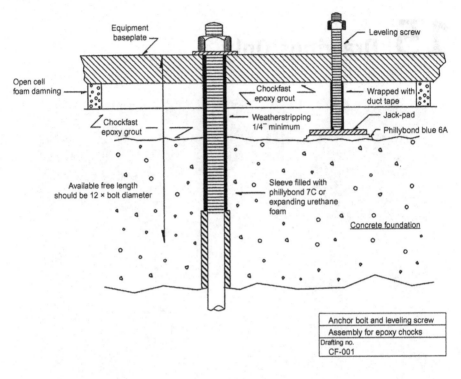

Equipment baseplate

Leveling screw

Open cell foam damning

Chockfast epoxy grout

Wrapped with duct tape

Chockfast epoxy grout

Weatherstripping 1/4⁻ minimum

Jack-pad

Phillybond blue 6A

Available free length should be 12 × bolt diameter

Sleeve filled with phillybond 7C or expanding urethane foam

Concrete foundation

| Anchor bolt and leveling screw |
| Assembly for epoxy chocks |
| Drafting no. CF-001 |

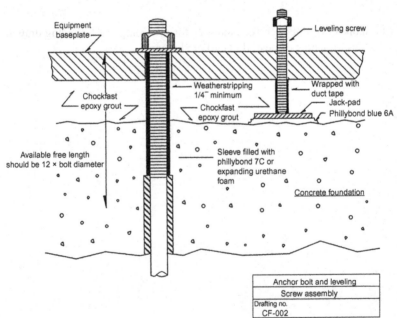

Equipment baseplate

Leveling screw

Chockfast epoxy grout

Weatherstripping 1/4⁻ minimum

Wrapped with duct tape

Chockfast epoxy grout

Jack-pad

Phillybond blue 6A

Available free length should be 12 × bolt diameter

Sleeve filled with phillybond 7C or expanding urethane foam

Concrete foundation

| Anchor bolt and leveling |
| Screw assembly |
| Drafting no. CF-002 |

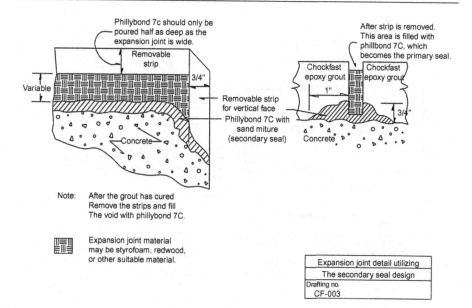

Phillybond 7c should only be poured half as deep as the expansion joint is wide.

Removable strip

3/4"

Variable

After strip is removed. This area is filled with phillbond 7C, which becomes the primary seal.

Chockfast epoxy grout

Chockfast epoxy grout

1"

3/4"

Removable strip for vertical face

Phillybond 7C with sand miture (secondary seal)

Concrete

Concrete

Note: After the grout has cured Remove the strips and fill The void with phillybond 7C.

Expansion joint material may be styrofoam, redwood, or other suitable material.

| Expansion joint detail utilizing |
| The secondary seal design |
| Drafting no. CF-003 |

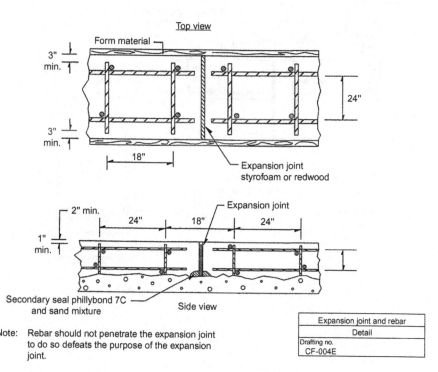

Top view

Form material

3" min.

3" min.

18"

24"

Expansion joint styrofoam or redwood

2" min.

24"

18"

24"

Expansion joint

1" min.

Secondary seal phillbond 7C and sand mixture

Side view

| Expansion joint and rebar |
| Detail |
| Drafting no. CF-004E |

Note: Rebar should not penetrate the expansion joint to do so defeats the purpose of the expansion joint.

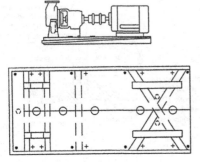

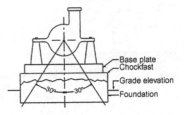

Base plate
Chockfast
Grade elevation
Foundation

Typical installation
Imaginary lines extended downward 30 to either
side of vertical and should pass through bottom
of foundation.

Foundation mass should be approximately 3–5 times
that of the equipment.

Pump base plan view

○   Existing grout holes.

◌   Indicates additional grout holes that may require field
    installation.

+   Indicates 1/2" vent holes that hay require field installation.
    additional vent holes may be required. Contact grout
    manufacturer.

•   Indicates possible vent holes installed by OEM.

--- Indicates cross bracing that may exist. But not be visible.

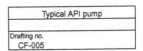

Typical API pump

| Drafting no. |
| CF-005 |

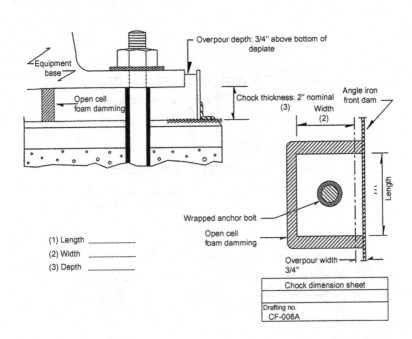

Overpour depth: 3/4" above bottom of
deplate

Equipment base

Open cell foam damming

Chock thickness: 2" nominal
(3)

Width
(2)

Angle iron
front dam

Wrapped anchor bolt

Open cell
foam damming

(1) Length _____

(2) Width _____

(3) Depth _____

Overpour width
3/4"

Length

Chock dimension sheet

| Drafting no. |
| CF-006A |

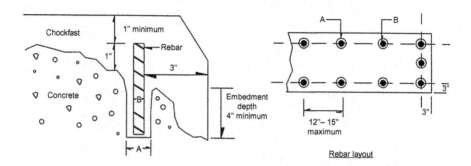

A - 1" inch larger than the rebar diameter

B - 1/2" inch diameter minimum

Dowels to prevent edge lifting
Drafting no.
CF-007B

References

Bickford, J. H. (1983). That initial preload, what happens to it? *Mechanical Engineering*, October.

Conversations with Bruce Shipley, P. E., Montgomeryville, PA.

Conversation with Perry, C., Monroe, P. E., Jr., Retired, Livingston, TX.

Conversations with W. E. (Ed.) Nelson, P.E., ASME, Turbo Machinery Consultant (deceased), Dickinson, TX.

Harrison, D. M. (1991). Chockfast machinery grouting manual, ITW Philadelphia Resins.

Harrison, D. M. (1992–1995). Papers presented to the Gas Compressor Short Course, University of Oklahoma, Norman, OK.

Harrison, D. M., Papers presented to the Process plant reliability conference, Houston, TX, November 1995 and October 1998.

Harrison, D. M. (2000). *The grouting handbook*. Houston, TX: Gulf Publishing Company.

Kauffmann, W. M., P. E. (1978). Analyze your engine foundations.

LOCTITE Corporation literature, Guide to designing assemblies with threaded fasteners, reprinted from an article appearing in *Mechanical Engineering*, October 1983.

Numerous papers, Dr. Anthony Smalley, Southwest Research Institute, San Antonio, TX.

Perry, C. Monroe Jr., P. E. (1989). Mounting pump baseplates. *Plant Engineering*.

Relationship of torque, tension and lubrication in threaded fasteners, FASTORQ Bolting Systems Inc., literature (date unknown), Houston, TX.

Sales and technical literature concerning engineered or load-indicating bolting systems from Rotabolt Ltd., West Midlands, England.

Technical literature from ITW Polymer Technologies, Chockfast, Escoweld, and CWC grouting products.

Technical literature, Master Builders Technologies.

Technical literature from Superbolt, Carnegie, PA, bulletin from June 1997.

Technical literature, US Grout, Five Star Epoxy Grout.

Printed in the United States
by Bookmasters